Springer Biographies

The books published in the Springer Biographies tell of the life and work of scholars, innovators, and pioneers in all fields of learning and throughout the ages. Prominent scientists and philosophers will feature, but so too will lesser known personalities whose significant contributions deserve greater recognition and whose remarkable life stories will stir and motivate readers. Authored by historians and other academic writers, the volumes describe and analyse the main achievements of their subjects in manner accessible to nonspecialists, interweaving these with salient aspects of the protagonists' personal lives. Autobiographies and memoirs also fall into the scope of the series.

More information about this series at http://www.springer.com/series/13617

Pieter C. van der Kruit

Master of Galactic Astronomy: A Biography of Jan Hendrik Oort

 Springer

Pieter C. van der Kruit
Emeritus Jacobus C. Kapteyn
distinguished professor of astronomy
Kapteyn Astronomical Institute
University of Groningen
Groningen, The Netherlands

ISSN 2365-0613 ISSN 2365-0621 (electronic)
Springer Biographies
ISBN 978-3-030-55550-4 ISBN 978-3-030-55548-1 (eBook)
https://doi.org/10.1007/978-3-030-55548-1

Cover picture: Jan Hendrik Oort in 1978, at the time of the commemoration of the fact that a hundred
years before that Jacobus C. Kapteyn, Oort's 'Inspiring Teacher', was appointed professor in Groningen.
The painting by Jan Pieter Veth shows Kapteyn at his desk. The box contains photographic plates, taken
at the Royal Cape Observatory, for the Cape Photographic Durchmusterung. To the right the plate holder
that Kapteyn used to measure the positions of almost half a million stars on these. Location: Kapteyn
Room at the Kapteyn Astronomical Institute, University of Groningen.

This Springer imprint is published by the registered company Springer Nature Switzerland AG
The registered company address is: Gewerbestrasse 11, 6330 Cham, Switzerland

Jan Hendrik Oort in 1973. From the Oort Archives

This book is dedicated to

Vera
Sylvia, Loes & Bo
Maurits, Karen & Dexter

and to
Jörgen & Xiang.

Preface

I tried to imagine the situation of a historian of natural sciences,
working some time in the twenty-first century
on a monograph describing Jan Oort's scientific importance.
If successful, this historian would have covered in his monograph
a very substantial portion of the history of astronomy in our century.
Bengt Georg Daniel Strömgren (1908–1987)[1]

Jan Hendrik Oort (1900–1992) is widely regarded as one of the most prominent astronomers of the twentieth century, building on the work of his professor, inspirator and teacher Jacobus Cornelius Kapteyn. The European Space Agency ESA's GAIA spacecraft has attracted much attention; its work of elucidating the structure of our Milky Way Galaxy builds directly on the fundamental work of Kapteyn and Oort. The present volume, together with a similar one on Kapteyn, traces among others the developing insight in this important branch of astronomy during the twentieth century.

No comprehensive, scientific monograph on Oort's life and work had been attempted in the first two and a half decades after his death in 1992. In August 2019, I published the volume *Jan Hendrik Oort: Master of the Galactic System* in the Springer Astrophysics and Space Science Library [2]. That work was written for a scientific audience. It is a sequel to my biography of Jacobus Cornelius Kapteyn (1851–1922), which had appeared in the same series in

---

[1]The quote from Strömgren comes from the *Liber Amicorum* written for Oort's eightieth birthday (*LibAm80*) [1].

2014, *Jacobus Cornelius Kapteyn: Born investigator of the Heavens* [3]. The current volume aims at a wider audience with a general interest in science. It is a sequel to the biography of Kapteyn, *Pioneer of Galactic Astronomy; A biography of astronomer Jacobus C. Kapteyn* [4] published together with the present volume in the series Springer Biographies. It has been published first in Dutch [5] and this is an English translation by the author.

As Bengt Strömgren notes in the quote above, Oort influenced most of the astronomical research in his lifetime. Chapter 2 describes the state of affairs regarding Galactic structure—and Kapteyn's contributions to that—around 1920, when Oort performed his first research as part of his studies with Kapteyn in Groningen. At other places, where appropriate, I have summarized what was known of the subject at the time.

Sources. A number of important publications related to Oort are references [1] through [10], see Appendix C. In particular, Jet Katgert-Merkelijn's extensive inventory of the Oort Archives [7] has been of enormous value. An important source for writing the biography are after all these Oort Archives, located at the Leiden University Library. I have spent about 75 days there going through this very extensive collection. Especially the private parts, which are only accessible with permission from the Oort family, have produced a lot of useful material. Oort was not only someone who kept everything (up to small pieces of paper in which he noted down his thoughts), but also someone who, by means of notes in diaries and pocket diaries, recorded what the daily experiences and events were. These and his letters to family members and colleagues have been invaluable.

I have used many **direct quotations** from letters, diaries and other notes in the Oort Archives to preserve the authenticity. Rewording or paraphrasing them would have taken away from the personal flavor and have given less insight into the personality of Oort.

Website. As in the case of Kapteyn, I have created a special Website about Jan Hendrik Oort: www.astro.rug.nl/JHOort. There one can find not only links to many publications and other material about Oort, but also to his own publications. Also, a large part of his archives are made available. My modus operandi in studying the archives was to take a picture on my tablet (an iPad) of every piece of paper that seemed worth keeping for later reference. In the end, I collected in this way a little over 30,000 pictures of which 10% or so turned out to be of insufficient quality. I have kept a complete archive of these original recordings. I reduced each usable picture by a factor of 4 (2 by 2 pixels), and collected them in pdf-files per catalogue number in the inventory of Jet Katgert. Except of course for the private sections, all these files can be downloaded from my Oort Web page.

Furthermore, Jet Katgert also gave me access to her archive of scanned photographs from the Oort collection. A special website with mainly group photos on which Oort appears, is available at www.strw.leidenuniv.nl/oortfotos/. I also have received a great deal of material from Oort's younger son Abraham Hans (Bram) Oort in the form of documents and photographs; this material is not publicly accessible.

The Website contains also a list of Oort's publications with links to the Astrophysics Data System, where electronic copies of many of his papers can be found. The Website also has scans of his popular articles such as in the amateur astronomer magazines *Hemel & Dampking* and *Zenit*, but these are in Dutch, but in many cases English translations have been provided. In the appendices of the Kapteyn Legacy Symposium [11] and the ASSL Oort biography [2], I have provided English translations or transcripts of important lectures that Oort gave at special occasions.

Numbers between square brackets refer to literature sources or to Websites in the section References. Comments on these are in footnotes, but I tried to keep these to an absolute minimum not to interrupt the reading of the main text unnecessarily.

I will not use any **nicknames** as they were used in the family. I do use first names as were widely used. For example, Oort's younger brother Arend Joan Petrus was called 'John' by his friends and acquaintances. I will use that too, but not the name 'Jo', which family members gave him. Likewise, I will use the name 'Hein' for Oort's older brother. Henricus, because he was addressed as such by many. The younger son, will be denoted as Bram Oort.

As a student and Ph.D. student with Oort as supervisor, I was addressed by Oort as 'van der Kruit', but from the day of my thesis defense he used my first name 'Piet'. I kept addressing him orally or in correspondence as 'Professor Oort'. In the 1980s, he and his wife suggested to me that from now on they should be called 'Jan' and 'Mieke', but I have hardly done so. In this book, I refer to them as 'Oort' and 'Mrs. Oort'.

In the text, when persons are mentioned for the first time, I give their full first names and (maiden) surnames, years of birth and—if applicable—death. The corresponding page then is the first entry for them in the Index.

Genealogy. There are a number of useful Websites for family tree research, certificates or family messages. The ones I have used are [12] to [18].

Currency calculators. To convert currencies from other times or countries into (at least indicatively) Euro's of the present time I have used Websites [19] to [21].

Geographical locations In Fig. 1.1, I provide a map of the Netherlands with the location of most places that are mentioned in this book indicated, especially the ones where Oort lived and worked.

It was a special privilege to have the opportunity to write a biography of Jan Hendrik Oort, because I am in the fortunate position to have known him over a long period of time. As a Leiden student I attended his lectures; for my candidates (bachelors) about the solar system and for my doctoral (masters) degree in stellar dynamics, respectively, in the academic years 1964–65 and 1966–67. After my bachelors in 1966 I came to work at the Observatory, where Oort was director, as an undergraduate student. I did my masters thesis under his supervision, part of the requirements of which were two research projects. The first concerned high-velocity clouds and inflow of intergalactic gas into the Galaxy; the second a Dwingeloo observation program to neutral hydrogen gas around the direction toward the Galactic Center. The last one became eventually, after my masters degree in 1968, my Ph.D. thesis research subject, on which I wrote a thesis with Oort as my 'promotor' (supervisor), which I defended in 1971. From 1970 on, I worked with the new Westerbork Synthesis Radio Telescope, that then had just been put into operation. Oort had a keen interest in that work, and in one case, he became co-author of the publication. Mainly thanks to an extremely positive recommendation by Oort I was offered a prestigious Carnegie Fellowship at the Mount Wilson and

Fig. 0.1 Jan Hendrik Oort and myself at the reception after my inaugural lecture as professor of astronomy at the University of Groningen in January 1988. From the author's collection.

Palomar Observatories in Pasadena, California, where I moved to in August 1972. After my return to the Netherlands in 1975, I worked at the University of Groningen, where I was appointed at the Kapteyn Astronomical Institute. In 1987 I was promoted to full professor. Oort kept a keen interest in my work. In January 1988, he attended my inaugural lecture in Groningen (see Fig. 0.1). He was a source of inspiration for me and a warm person with much personal interest, an inspiring teacher and later colleague. I have tried despite all this to make this biography as objective as possible.

I hope this book and the accompanying Website with links to other sources and publications will enhance the appreciation for Oort as a human and as an astronomer. The story of Oort's scientific developing insight and its significance follows directly on to that of his mentor and inspiring professor. In that way, this book is a sequel to my biography of Oort's teacher Jacobus Cornelius Kapteyn. Their work finds at present a fascinating culmination in the fundamental discoveries of ESA's astrometric satellite GAIA. I hope these two biographies together form a tribute and does justice to these two great scientists in whose tradition I have had the privilege of conducting my own astronomical research.

Groningen, The Netherlands Pieter C. van der Kruit
Original February 2019–September 2019, Emeritus Jacobus C. Kapteyn
Translation November 2019–May 2020. distinguished professor of astronomy
Kapteyn Astronomical Institute
University of Groningen
https://www.astro.rug.nl/~vdkruit

For more background and information on Oort please consult my scientific biography [2] in the Astrophysics and Space Science Library and the dedicated Website www.astro.rug.nl/JHOort.

Acknowledgments

Many people have made crucial contributions to the success of this project to write a biography of Jan Hendrik Oort. The Springer Astrophysics and Space Science Library version [2] has a more comprehensive summary and I will limit myself somewhat here.

First of all, I would like to thank Bram Oort, younger son of Oort, and Marc Oort, grandson of Oort who has a Ph.D. in astronomy. They not only provided much material and answered questions, but they also commented extensively on the manuscripts of both the extended version and of the Dutch version on which this translation is based. Jet Katgert-Merkelijn has done an enormous amount of work organizing the Oort Archives (together with Oorts long-time secretary, Dinie Ondei-Beneker) and producing an inventory of the Archives, which has saved me several years of work. Just after this manuscript was submitted to Springer and before it went into production we sadly learned of the untimely death of Jet at the age of 76 years as a result of an unfortunate and fatal fall in her home. This Oort biography and the academic one together form a tribute to her dedicated and painstaking work.

A number of 'proofreaders' (not in the sense of proofs of printed texts, but of the Dutch 'wijnproeven', wine tasting) have commented on the manuscript of the original scientific biography. Jet Katgert and my colleagues Butler Burton and Jan Willem Pel have carefully looked at the astronomical aspects in particular, and historians Klaas van Berkel and David Baneke at historical matters. Also, my good friends Albert Jan Scheffer and Ton Schoot Uiterkamp have read and commented on the manuscript of the original book. I also thank the proofreaders, who read the manuscript of the Dutch book for a wider audience [5] and commented very helpfully on it. These are my wife Corry, and my very helpful good friends Geert Hoornveld and Rob Knol. The work of all my proofreaders has been accurate and excellent; however, errors that have been left behind are entirely my own responsibility.

I am also very grateful for the very pleasant help and support of the Leiden University Library, Department of Special Collections during my 75 or so days of going through the Oort Archives. I am also grateful to the staff of the Archives at the California Institute of Technology, Pasadena, California and the Huntington Library, San Marino, California, for help when I examined correspondence with Oort from local astronomers.

For all reproductions, I have been granted permission for use in this book by the holders of the copyrights. I am grateful for their help, particularly the Kapteyn Astronomical Institute and the Sterrewacht Leiden.

Many thanks are also due to my editors of the previous versions, Ramon Khanna of Springer and Job Lisman of Publishing house Prometheus.

I am also very grateful to Springer and editor Ramon Khanna for their willingness and help to publish this biography and the accompanying one on Oort's 'great mentor' Jacobus Cornelius Kapteyn. The translating from the original Dutch version has profited enormously from the availability of DeepL Translator Pro version (www.deepl.com/), which provided an excellent first step in translating the Dutch manuscript.

I would like to thank the Kapteyn Institute, my colleagues (astronomers and staff of the secretariat and of the computer group) for their help and interest. I am grateful to the directors, Reynier Peletier and Scott Trager for their hospitality to generously accommodate an emeritus professor. My work on an Oort biography has been financed and/or otherwise supported by a number of organizations. On the recommendation of the Faculty of Mathematics and Natural Sciences, the Board of the University of Groningen appointed me in 2003 on the distinguished Jacobus C. Kapteyn chair in astronomy. The faculty accompanied that, until my formal retirement in 2009, with a generous annual allowance to be spent on research of my own choosing. The part of it that was left over, has been used for this project. Furthermore, I am grateful for the support of the Kapteyn Astronomical Institute in Groningen, Leiden Observatory, the Dutch Research School for Astronomy (NOVA), the directorate Exact Sciences of the Netherlands Organization for Scientific Research (EW-NWO) and the Jan Hendrik Oort Fund.

And finally I thank my wife Corry for her love, understanding and support, with which she made it possible for me to write this book.

Contents

Fig. 1.0 Map of the Netherlands with places where Oort lived and worked (capitals): F=Franeker, where he was born, O=Oegstgeest, where he grew up, K=Katwijk aan zee, where his parents had a second home, L=Leiden, where he went to school and worked most of his career, G=Groningen, where he studied at the university, N=Nunspeet/Hulshorst, where he lived during most of WWII, H=Haamstede, where the Oorts had a second home, W=Wassenaar, where he lived his last years and died, and the radio observatories at Ko=Kootwijk, D=Dwingeloo and We=Westerbork. Other locations with two-letter codes for reference: al=Alkmaar, am=Amsterdam, an=Antwerpen, ar=Arnhem, br=Breda, do=Dordrecht, dh=Den Haag, dv=Deventer, eh=Eindhoven, en=Enschede, ha=Haarlem, hd=Harderwijk, hr=Harlingen, le=Leeuwarden, ma=Maastricht, nij=Nijmegen, ro=Rotterdam, ut=Utrecht, vl=Vlissingen, ve=Velzen, wa=Wageningen, zu=Zutphen, zw=Zwolle. In 1900, when Oort was born, some features did not exist, notably the polders of Flevoland in the center; this was the *Zuiderzee* until it was closed by the *Afsluitdijk* (Closure Dam) in 1932, and became the *IJsselmeer*. Adapted from 'Kaart Nederland Jan clip art' [22]

1

Youth in Oegstgeest and Leiden

There is no elevator to success. You have to take the stairs.
Anonymous

There are no parents that do not secretly see their children as special.
And they are right, the possibilities are unlimited.
Godfried Jan Arnold Bomans (1913–1971)

The date is June 24, 1970. A helicopter lands on the grounds of Woonoord Schattenberg, so named after a nearby burial mound, about 7 km south of the city of Assen in the province of Drenthe (see Fig. 1.1). Since 1951, former soldiers of the Royal Netherlands-Indies Army KNIL, originally inhabitants of the Moluccas archipelago—now part of Indonesia—, had been accommodated there. They were awaiting (in vain) their return to their independent republic of the South Moluccas, which was proclaimed on the island of Ambon. In 1939 the complex of barracks, in which the Mollucans lived, was built as the Westerbork Central Refugee Camp—the name derived from a nearby village—for Jewish refugees from Germany and Austria. They were considered unwanted aliens and would return as soon as possible. In 1942, during the German occupation, the complex had been transformed into the Westerbork Concentration Camp, a transit camp for the deportation of eventually more than 100,000 Dutch Jews and a small number of Roma and other minorities to extermination camps in Germany, Poland and the Czech Republic. After

---

Godfried Bomans was a Dutch writer and television personality.

Fig. 1.1 Queen Juliana of the Netherlands landing at the site of the Westerbork Synthesis Radio Telescope in 1970. From the Oort Archives

the war it was used for a short time as an internment camp for collaborators with the Nazi regime.

A place with a rather unedifying history. The helicopter was occupied by Queen Juliana. Just a few months earlier, on May 4, 1970—on the eve of the 25th anniversary of the liberation of the Netherlands—, she had revealed the National Monument Westerbork. She now returned to inaugurate the most advanced radio telescope in the world. It was an impressive sight. Twelve dish antennas of 25 m diameter very accurately positioned on an east-west line of about one and a half kilometers length; from the west ten dishes exactly 144 m apart and a 300 m track on the eastern side with two more movable ones. This instrument would be the most sensitive in the world in its class for ten years, and continued to be the leading facility until the United States built an even bigger version, the Very Large Array near Socorro, New Mexico. Even then the Westerbork Synthesis Radio Telescope, which by then had been extended with two more dishes on a railroad track another kilometer and a half to the east, remained competitive. The Queen was met by Jan Hendrik Oort, professor of astronomy at Leiden, director of the Sterrewacht (Leiden Observatory),[1] and chairman of the Foundation Radio Radiation from Sun and Milky Way. He was the founding father of the new radio telescope. A few months earlier, Oort had turned seventy years of age and at the start of the new academic year he was to retire from these positions. He would still devote many years to astronomical research. For decades, he had been one of the world's leading, most important and authoritative astronomers, particularly in the field of the study of our and other galaxies, building on the groundbreaking work of Jacobus Cornelius Kapteyn, his 'great mentor' as he called him in his inaugural speech in 1936. How was it possible that a small country like the Netherlands was able to build such a unique instrument and take a big lead on the rest of the world in a branch of science like radio astronomy? How could it be that at the end of the twentieth century the Netherlands had become one of the most influential countries in the world in astronomy? To understand that, we need to go back to 1900, the last year of the nineteenth century, the year in which Jan Hendrik Oort was born.

1.1   Franeker

At the end of the nineteenth century, the Frisian city of Franeker, or in Frisian Frjentsjer, did not have more than six or seven thousand inhabitants. The town had been granted city rights in 1374 and was one of the eleven cities in Friesland on the route of the famous, about 200 km long ice-skating trip. When under the leadership of Willem van Oranje (1533–1584) the liberation struggle against the King of Spain had started, Franeker was the first of the Frisian cities to take his side. The new Republic of the Seven United Provinces had no universities; the only one in the former Netherlands, that of Leuven, fell outside the areas covered by the provinces that signed the Unie van Utrecht (Union of Utrecht) of 1579 and the Acte van Verlatinghe (Act of Abjuration) of 1581. Each of the provinces had the right to found a university and in anticipation of the developments the University of Leiden was established in 1575 (incidentally, under the authority of the King of Spain, Philips II!). Friesland followed as soon as possible and Franeker was rewarded for the early choice to the support the 'Father of the Fatherland'. The university was established there in 1584 as the second of the Netherlands.

The University of Franeker became a thriving center of science, especially in the seventeenth and eighteenth centuries. Lectures were given in theology, medicine, literature, philosophy, and mathematics and natural sciences.

---

[1]Throughout this book I will use the name 'Sterrewacht' for Leiden Observatory.

Famous professors include Petrus Camper (1722–1789), a physician, anatomist, zoologist, anthropologist, paleontologist and philosopher. He occupied the chair in philosophy from 1749 to 1755. Well-known students include René Descartes (1596–1650), who studied there between 1628 and 1630, and Peter Stuyvesant (c.1611–1672). Astronomy was taught by professors who were well-known at the time, Adriaan Metius (1571–1635), Johannes Phocylides Holwarda (1618–1651) and Jean Henri van Swinden (1746–1823). Johann Samuel König (1712–1757), student of the famous mathematicians Johann (1667–1748) and Daniel Bernoulli (1700–1782) in Basel, was professor of philosophy and mathematics in Franeker from 1744 until 1749. König is part of the academic genealogy (who is a person's PhD student or equivalent) of Jan Hendrik Oort.

In Franeker wool carder Eise Jelteszn Eisinga (1744–1828) built between 1774 and 1781 an orrery in his living room, a scale model of the Solar System with the then known planets. It is on a scale of one in a trillion (10^{12}) and still runs accurately; it is now a well-known tourist attraction of Franeker. Eisinga was also honorary (=unpaid) professor at the university.

In the course of the eighteenth century the number of students in Franeker decreased. The main reason for this was the competition from the University of Groningen, which was established as the third Dutch university in 1614. Petrus Camper was after his departure from Franeker for some time associated with the *Athenæum Illustre*, the precursor of the (Municipal) University of Amsterdam, before in 1763 he became professor in Groningen, where he worked for ten years. The end of the University of Franeker came with the French occupation. Napoleon and the French closed all Dutch universities (there were five by then) with the exception of those of Leiden and Groningen. When the Kingdom of the Netherlands was proclaimed in 1815, only the university of Utrecht, which had been founded in 1636, was allowed to open its doors again, but those of Franeker and Harderwijk remained closed. The Athenæum, that remained in Franeker, did not survive very long and was closed in 1843.

The question was of course what to do with the university buildings. The States of Friesland decided to use it as an asylum for lunatics. Until then, the mentally handicapped of Friesland were nursed in Deventer and that was a hundred km or so away. The first patients appeared in 1851 (27 in number). During the day they worked on nearby farms. In 1897 a new second physician was appointed to the asylum, by the name of Abraham Hermanus Oort (1869–1941).

Abraham Oort was born on 28 March 1869, also in Friesland, in the city of Harlingen, only 7 km from Franeker. He was a descendant of a long line of theologians and preachers. His father, Henricus Oort (1836–1927), had studied theology in Leiden and had obtained a PhD degree in 1860. That same year Henricus married Elisabeth Wilhelmina de Goeje (1838–1907) and became a preacher in Santpoort near Haarlem. Eventually, they would have eleven children. In 1866, Henricus was appointed minister at Harlingen, but would stay there only six years. In that period, Abraham Hermanus was born as the seventh child. The Oort family then moved to Amsterdam, where Henricus Oort worked for some time at the *Athenæum Illustre*, before he was appointed professor at Leiden University (in the Faculty of Arts). His chair was in Hebrew, ancient Israel and the Old Testament.

Henricus Oort was very closely involved in the publication of a Bible for Children, together with brother-in-law Isaäc Hooykaas (1837–1894), husband of Henricus' sister Petronella Everharda Oort (1839–1916), and his brother Abraham Johannes Oort (1838–1917), both of which were also preachers. This children's Bible has been translated into French and English, so it made a big impression. His other important work was a new translation directly from the Greek of the Old Testament, 'according to strict scientific principles', together with Hooykaas and the Leiden professor of theology Abraham Kuenen (1828–1891). They were among the liberal protestants. However, this 'Leiden translation' was not widely read outside Protestant circles. Henricus Oort also translated the New Testament.

The son Abraham Oort was six years old in 1875, when they moved to Leiden. He attended the primary school and the gymnasium, of which he took the final exam in 1887. Against the family tradition (and unlike two of his elder brothers), he chose to study medicine rather than theology. After his exam for physician in 1896 he decided to volunteer in a psychiatric clinic in Heidelberg [23]. The following year he married Ruth Hannah[2] Faber (1869–1957) in Leiden. Shortly thereafter, he accepted the position of second physician at the asylum in Franeker. Ruth Faber was born on 6 October 1869 in Ossendrecht, as the daughter of Jan Faber (1838–1889) and Henriëtte Sophia Susanna Schaaij (1840–1908). Faber also came from a tradition of preachers. He died just before his fifty-first birthday (and also before his daughter Ruth married Abraham Oort). At the time of his death he preached in the Reformed Church in Heemskerk near Haarlem, where he had been appointed a year earlier.

The young couple Oort–Faber (Fig. 1.2) settled in Franeker; the address was Zilverstraat 16 (see Fig. 1.3). This house is close to Eisinga's orrery and now contains a plaque, that marks it as the house where Jan Hendrik Oort, the subject of this biography, was born on 28 April 1900 (Fig. 1.4). He was preceded by his older brother Henricus (1898–1922), who was known as Hein

---

[2]Aficionados of linguistic curiosities (like myself) will notice that 'Hannah' is a palindrome. It is the longest palindrome among Dutch girl's names. For boys, that is Reinier [24].

Fig. 1.2 Abraham Hermanus Oort and Ruth Hannah Oort–Faber, father and mother of Jan Hendrik Oort, in 1902. From the Oort Archives

Fig. 1.3 Postcard with the birthplace of Oort. It's the house on the far right. In the middle the 'Koorndragershuisje', the grain carriers house. The card was not dated, but we see here the relevant part of Franeker as what it must have looked like when Oort was born there

(he lived for only 24 years due to diabetes). His younger brother and two sisters were born after the family had left Franeker. The birth certificate of Oort is reproduced in Fig. 1.5. Oort was born at 8:30 in the morning. The witnesses (at the registration) were Willem Ferwerda and Adriaan van Voorthuijsen. Ferwerda is described as an 'accountant' and was 41 years old. The only person of that name and born about 1859 and still living in Franeker (at least in 1888),

Fig. 1.4 Plaque to the facade of the house where Oort was born in Franeker. It is secured to the house to the right of the windows on the ground floor (see Fig. 1.3). Photograph by the author

which I could find, was a candidate civil-law notary. Adriaan van Voorthuijsen (1872–1952) was a physician/general practitioner in Franeker. He had studied medicine in Leiden and as Abraham Oort and van Voorhuijsen differ only by three years in age, they would have known each other from their university days. Van Voorthuijsen later moved to Groningen, where Oort often visited him when he was a student there.

1.2 Oegstgeest

The Oort–Faber family did not stay in Franeker for very long. Therefore, Jan Hendrik Oort was not much attached to Franeker and Friesland, probably just like his father, who was also born in Friesland and also moved to the West after a few years. Abraham Oort had kept in touch with his teacher in Heidelberg, Emil Kraepelin (1856–1926), who focused on experiments with persons while they performed concentrated mental work, studying the influence on their work of coffee, tea, alcohol, etc. He had collected, among other things, material on children of various ages and of adults doing mental arithmetic. Abraham Oort was allowed to use the data on the speed with which children added numbers,

Fig. 1.5 Birth certificate of Jan Hendrik Oort [25]

and studied the effects of age, social class, fatigue, etc. on their performance. He wrote a dissertation on this subject, *Experiments with continuous mental work by school children.* He obtained his doctorate in Leiden on November 26, 1900. His supervisor was the first professor of psychiatry in the Netherlands, Gerbrandus Jelgersma (1859–1942). He had studied medicine in Amsterdam and had developed an interest in the brain and the nervous system. He was appointed in 1899 and Abraham Oort was his first PhD student; Oort was awarded the doctoral degree *cum laude.*

Now, at Jergerma's appointment, the Municipality of Leiden had agreed that there would be a new sanatorium 'Rhijngeest' located in nearby Oegstgeest. It would be a clinic for the temporary treatment of patients with neurotic disorders. The Rhijngeest estate was owned by the municipality of Leiden, which had bought it in 1895 together with the nearby castle Endegeest, a medieval castle surrounded by garden and forest. The French philosopher René Descartes, whom we have met above, had lived there between 1641 and 1643. The city of Leiden wanted to establish an asylum for lunatics in Endegeest, and did so in 1897, even before the appointment of Jelgersma. Endegeest is still in use in mental health care. This is something else than the Rhijngeest clinic on

Fig. 1.6 Photograph probably taken in 1902 with Oort on the left and his older brother Hein on the right. It was not unusual that little boys like Oort here up to the age of two years or so were dressed this way

the other side of the street, which by the way, after later having been called the Jelgersma Clinic, was dissolved in the 1990s and now houses the town hall of Oegstgeest. Although Rhijngeest was not part of Leiden University, it was part of the agreement that brought Jelgersma to Leiden. Jelgersma was unable to run the Rhijngeest sanatorium next to being professor at the university. The City Council of Leiden decided in December 1902 to appoint Dr. Abraham Oort first physician and director of the Rhijngeest sanatorium [26]. There is no doubt that Jelgersma must have had a hand in that. The Oort–Faber family moved in 1903 to Oegstgeest, when Jan Oort was less than three years old (see Fig. 1.6).

The director of Rhijngeest lived in the official residence on the grounds of the sanatorium (Fig. 1.7) and this is the place where Jan Hendrik grew up. Here

Fig. 1.7 The house of the director of Rhijngeest, where Oort grew up. From Wikimedia Commons [27]

the younger brother and sisters of Oort were also born: Arend Joan Peter on 27 May 1903, Jetske Sophia Susanna on 24 June 1904 and Emilie Annette on 6 March 1908. Arend Joan Peter was usually called 'John', but among friends and in the family was 'Jo' (pronounced 'Yo'). Jetske was in family circles 'Jeppie' and Emilie 'Emy'. I will not use these nicknames, but will use the more commonly used names Hein for the older brother Henricus, and John. The Oorts also owned a second (vacation) home in Katwijk aan Zee, near Leiden at the coast, which was called Sandy-Hook, and where weekends and summer holidays were spent.

Abraham Oort was not affiliated with the university, but he nevertheless did other psychiatric work besides running the sanatorium. He developed tests for the admission exams for nurses who wanted to specialize for psychiatric institutions. Oort described his father in an interview[3] as very 'much interested in science. He made some simple brain experiments and was one of the first in the Netherlands to use psychological tests.' He also has for a long time provided every five years a revised edition of a textbook for psychiatry. This book, *Lectures on the care of patients that are mentally ill and lunatics*, was published in 1906 by Dr. Jan Christiaan Theodoor Scheffer (1855–1905), who

---

[3]This is an interview for the American Institute of Physics in 1977 and is further referred to as the *AIP-Interview* [6].

Fig. 1.8 Oort with its brothers and the elder of his sisters, probably in 1908 or somewhat earlier. Brother Hein is at the left and brother John at the bottom-right next to Sister Jetske. From the Oort Archives

had been the director-physician at Endegeest. He died young, even before the book appeared, and Abraham Oort saw it as his duty to keep it up to date. He was also closely involved in an organization of physicians who opposed the use of alcohol. They considered addiction to be a disease, which—like the then commonplace tuberculosis and syphilis—had to be treated. Oort's thesis had been accompanied by a proposition, saying that new institutions had to be set up in such a way to admit alcoholics, if necessary involuntarily. Abraham Oort and his wife were teetotalers all their lifes. Jan Oort has not followed this example.

The following description of the grandparents and their household at Rhijngeest was made available to me by Jetske (Jornaroos) Hamaker-Nauta (b. 1936), daughter of Oort's sister Emilie Annette (Figs. 1.8 and 1.9).

Fig. 1.9 Oort (right) and his older brother Hein around 1908 or a little later. From the Oort Archives

... On the left was the house of the physician-director, [...] and there was also a villa where Prof. Suringar, the botanist, had lived, who had grown many interesting plants in the park. Some private patients were living there as well. The house was high, cold and small for a family of 5 children and an intern handmaiden, who still had to address the little girls Miss Jetske and Miss Emy. There were only two washbasins in the house, you had to wash yourself in the bedrooms in bowls with a jug of water put next to it. Sometimes in winter there was ice in them. Only on Saturdays the geyser was turned on so that one could wash oneself with a thick beam of hot water. [...]

Of course, the park lent itself to endless wandering and discoveries, in the burrows under the rhododendron bushes and in the ponds. Ponds with back-swimmers, water beetles and sticklebacks and salamanders. And sometimes there was a kingfisher in 'Hudson Bay' [...] and there was also a large 'Belle de Boskoop' [a kind of apple] tree. In early fall, the apples were carefully picked with a kind of net on a stick and then stored in the attic and turned every week, only if there were any spots on them they were allowed to be eaten: the ones in good order were only eaten at Christmas.

[And then there were] the chickens. Sometimes you had to put aside hands full of stellaria and pick young urtica before you could reach them. [...] and then they would pick up the eggs, wash them and date them. Eggs were never bought. The young roosters were swapped with head-sister Dunnebier; you would not put your own cockerels in the pan so easily. This nicest nun, still dressed in habit, also supplied the broody hens, so that chicks would come. Sometimes they were hatched under a lamp and helped with the breaking of the eggs. [...] The mother was an animal lover, she imitated, by tapping with her nail, how the mother hen taught her chicks to pick seeds by tapping her beak, and the chicks understood what she meant. In the family, she was called 'Grandma Chickens'. She also kept bees, she would put a cap on her head and a smoking pipe in her mouth. The honey was centrifuged by themselves. **[...]**

The mother was a red woman [had socialist sympathies], and she approved very much of the idea of Co-operations, buying together, no profit required. Once a month a big box with all the groceries arrived. [...] In the family there was a great fear of tuberculosis. Understandable, for 8 of the 16 children in the father's family had died of it, usually because of the complication meningitis. Father's mother had also suffered from this disease herself. So the father was terrified of all infectious diseases and infections.

And about her grandfather, Oort's father:

He was physician-director of a psychiatric hospital in Oegstgeest, but you would not tell if you met him; a small, shy, very modest man, quiet and perhaps almost boring, with small very friendly eyes, an alpine cap and a small white beard. He was not a man of many words: telling stories, no, I did not learn that from him. He had stomach problems at some point and had to spend a long period of time resting and in that time he started wood carving with a small knife and with great precision he cut a wooden box for all the grandchildren with their name on it.

Oort's younger son Abraham Hans (Bram) Oort wrote:

Jan's parents both came from a long family of vicars and theologians. Grandma Chickens and her eldest daughter Jetske went regularly to the Remonstrant Church in Leiden. The other children Jan, Jo, and Emy were not religious, rather at odds with this family tendency. Grandma told us that Jan in particular was always obstinate; not only about religion in general, they called him Jantje contramine. Jan was a real scientist without faith in a God but with a feeling about some higher power in nature. He loved to walk in the unspoiled nature; 'no entry'-signs attracted him to enter and find out what it was behind it that was forbidden. I believe he has never been fined and he always could talk his way out of it in a friendly way when he was caught!

Grandma Chickens was a dear, soft grandmother but she could be moralizing. [...] Grandma often told me about her long voyage by (sailing?) ship as a young girl to Indonesia. She remembered the great impression of the first sight of the Cape of Good Hope. The Faber family went for some years to Indonesia for a new preacher's job her father had accepted. Her younger brother, Paul Faber, was a baby at the time. There was a cow on board for milk for the babies on the boat. Later, Uncle Paul was my favorite uncle; he was retired at the age of 50 having been a member of the Supreme Court in the Dutch East Indies; and lived deep into his 90s.

Paulus Frederik Karel Faber (1874–1973) lived to 98; much older than his wife Clara Cornelia Steynis (1880–1937).

Grandma Oort did social work in the slums of Leiden often accompanied by Jetske and Emy. There was a lot of poverty and a lot of alcohol problems in Leiden. Jetske and Emy both went into social work. Jetske remained unmarried and worked as a nurse; she was the one who cared the most for Grandma Oort when she got older. Emy married Jelle Nauta, a competent radiologist in Rotterdam.

Jetske became director of the Student Sanatorium in Laren. The radiologist was Jelle Haring Nauta (1907–1987).

1.3   'Hoogere Burgerschool' in Leiden

Jan Hendrik Oort attended the (public) primary school in Leiden without any problems. His grades were excellent: on a scale from 1 to 10, between 7 and 9 for all subjects and for things like behavior, homework, etc., except for carefulness, for which he scored a 10. Father Oort also had a strong interest in botany and other science. The two oldest boys were more interested in other subjects, but the younger brother looked more like the father in that respect. Eventually, the older two studied physics and astronomy and the younger one biology. This dichotomy was already noticeable among the three boys from their choice of secondary education: Hein and Jan went to the 'HBS', John to the gymnasium (grammar school). This choice of school deserves some explanation.

De 'Hoogere Burgerschool' or HBS had been established in 1863 by Johan Rudolph Thorbecke (1798–1872), the liberal statesman who had also been responsible for the new Constitution of 1848. As Prime Minister, he had modernized the educational system, first the primary schools and then the secondary education by introducing the HBS. This was meant to be a high-school education for boys from the well-to-do bourgeoisie preparing them for a career in commerce and industry. Greek and Latin was therefore not in the curriculum, but much attention was paid to modern languages and science. The level of education in the natural sciences was very high; teachers in mathematics, physics and chemistry were academics with PhD's. It is believed to have been an important factor in the relatively large number of Nobel prizes for physics and chemistry by Dutch scientists in the first years these were awarded ([28], English abstract in [29]): Henricus van 't Hoff (1852–1911), Hendrik Antoon Lorentz (1853–1928), Pieter Zeeman (1865–1943) and Heike Kamerlingh Onnes (1853–1926) had all gone through the HBS. Johannes Diderik van der Waals (1837–1923) had not even attended gymnasium or HBS; he had special permission from Thorbecke to enter university. The Prime Minister had several similar powers; one of Thorbecke's last decisions was to admit Aletta Henriëtta Jacobs (1854–1929) as the first woman to the University of Groningen to study medicine.

The HBS did not automatically allow one to enter university, not even for such studies for which knowledge of classical languages was not a necessity. There was a possibility to take a special entrance exam. That route had, for example, been taken by Kamerlingh Onnes [30]. He grew up in Groningen, where the first HBS was opened in 1864 (the municipal council had already submitted a request for its establishment to Thorbecke before the law was actually passed). Kamerlingh Onnes entered one year later. The entrance exam was apparently not very difficult, at least not for the University of Groningen. This route via HBS and such an exam was a common one. For medical students it was less attractive, because with an HBS diploma one could study medicine and become a physician, but one could not defend a PhD thesis and obtain a Doctor's degree. You had to go to a foreign university for that. For a selected few there was the possibility of being awarded a doctorate *honoris causa*, as in the case of Jelgersma at the University of Utrecht.

The law that regulated access to the university with an HBS degree was only passed in 1917, the year in which Oort obtained that diploma. But much earlier it had become common practice for boys with a predisposition for mathematics and science subjects to choose for a proper training in those subjects and enter the HBS. This was also the case in the Oort family, where the older brothers Hein and Jan went to the HBS (see Fig. 1.10) and the younger brother John with his interest in biology the Gymnasium.

Hein Oort went on to study physics in Leiden. About his younger brother John, Oort said in the *AIP-Interview* [6]:

> He [father Oort] was also very much interested in nature, especially botany and plants; those interests were inherited by my younger brother, who became a naturalist already from his school days. I think in a way he was much more pronounced in his direction than I was. My mother had undergone a severe operation and had to go to Italy for a recovery period after that. She got letters from the children. In letters from me arithmetical problems were sometimes scribbled in the margin. When she got a letter from my youngest brother it was always about flowers, birds and mushrooms he had found. In fact, when he was still in the lower grades he discovered a mushroom in the woods which had never

Fig. 1.10 Front of the building of the Hoogere Burgerschool Leiden. In the back we see part of the Pieterskerk, the main church building of Leiden, and now the location of many formal ceremonies of the University of Leiden. From www.Erfgoedleiden.nl [31]

been found before in Holland. He was at that age already quite an expert. He later became a professor of plant diseases at the University of Wageningen.

'Holland' was often used instead of 'the Netherlands', especially by contemporaries of Oort. It is done less often nowadays; Holland actually refers only to the two western provinces (containing Amsterdam, Den Haag and Rotterdam) and not the whole country. It is similar to the incorrect use of 'England' when the whole of Great Britain or UK is meant. John Oort eventually became professor of phytopathology in 1949 and was director of the famous Laboratory of Mycology (the study of mushrooms) and Potato Research, later the Laboratory of Phytopathology at the Agricultural University Wageningen [32]. The daughters Jetske and Emilie attended the HBS for girls, which by then existed.

Oort is most likely present as a fifteen-year-old in Fig. 1.11. I did send this picture together with an earlier and later photograph of Oort, to the Department of Artificial Intelligence at the University of Groningen, but I was informed that for facial recognition sharper images are required. So I have been conducting a survey among staff and students of the Kapteyn Astronomical Institute on the basis of eighteen selected faces from the photograph of boys of about Oort's age, along with the two pictures for comparison. The two most frequently chosen and one of the comparison photo's (on which Oort is twenty) have been reproduced on the right side of the figure. The most

Fig. 1.11 Teachers and pupils of the Hoogere Burgerschool Leiden in lieu of the opening of the new building in 1915. The older boys wear hats or caps, the younger ones are bareheaded. I conducted an identification survey among staff and student of the Kapteyn Astronomical Institute about who would be Oort (see text). On the right, the top two pictures are first and second choices in this poll. My own preference is the lower picture. At the bottom for comparison Oort at the age of twenty from Fig. 3.1. From www.Erfgoedleiden.nl [33]

frequently chosen option is the top one, but a close examination of this person in the group photograph indicates that, unlike Oort, he clearly was of well above average height. The second one was also my choice and is probably Oort as a fifteen-year-old.

The Oort Archives in Leiden contain in the private parts diaries that go back to his teens.[4] The oldest notes are from December 24, 1915. Unfortunately, there is little to be found about Oort's choice of field of study. During his high school time often a girl (Kitty) is mentioned, with whom he apparently had fallen in love. These notes show a characteristic of Dutch society, namely the pillarisation. The girl in question was of Catholic descent and her parents and the parish priest probably had forbidden her to have dealings with children from the other pillars (such as Protestants). Oort met her regularly at parties, sports games and the like, but she remained cool to him. The pillarisation was a feature of Dutch society since the end of the nineteenth century. The reform of the Constitution of 1848 and the reorganization of primary education under the leadership of Thorbecke would have improved the position of Catholics, for example Catholic elementary schools had been allowed. But the stricter Protestant circles resisted these liberal ideas, first under Guillaume (William) Groen van Prinsterer (1801–1876) and later Abraham Kuyper (1837–1920),

---

[4]The Oort Archives are located at the Special Collections department of Leiden University Library. On my Website www.astro.rug.nl/JHOort I provide access to my photographic recordings of the Archives. The private parts, like these diaries, however, are confidential and not included there.

which represented the more orthodox Protestants and opposed interaction with Catholics. There was also a third pillar, usually referred to as social-democratic. There remained atheists or those Catholics and Protestants, who had loose connections with their churches. Oort belonged to the last group.

From Oort's diaries, written on February 6, 1917:

> Great, healthy sport that skating. This afternoon I skated at high speed with Paul Reinhoed to Katwijk, where he wanted to show his grades. The weather was beautiful. The beach was full of big blocks of ice stacked in fairly high heaps on top of one another, which sometimes protruded into the sea. The water in between the ice heaps froze while flowing.
>
> I more and more like the simple, ordinary life these days. I believe that, in the long run, it is only there that the deep beauty resides.
>
> As I wrote before, I'm happy with all the beauty that I see. All I really care for is Dickens and some fairy tales, because I like the conviviality in them and I love Dickens' fairy tales, those fairy tales are so beautiful and so free from roughness and evil.

In 1917 Oort passed the HBS final exam with flying colors. His grades, on a scale from 1 to 10, for mathematics and science subjects were: 10 for mathematics and for mechanics, 9 for physics and 7 for chemistry. He scored only just 'sufficient' for the languages: 7 for Dutch and a 6 for French, English and German. Greek and Latin at a gymnasium might have been problematic for him. For drawing he had a 5 ('doubtful'), perhaps not surprising for those who attended his courses and saw him use the blackboard. For geometrical drawing he had a 7, but then for this one was allowed to use ruler and compass.

As a teenager, Oort developed an interest in astronomy. He once attributed this to the books of Jules Gabriel Verne (1818–1905). His books were available in translation and they aroused his interest, as they have done with many teenage boys and (perhaps also then, certainly later) girls. Friends of his parents lent him a telescope, which he used at night to prowl the sky. From the *AIP-Interview* [6]:

> But I certainly got interested in science and astronomy already in my high school years. I don't remember exactly where my first interest in the subject came from. I suppose it was by reading books by Jules Verne who was very popular in those days. *Around the world in 80 days, the voyage to the Moon* and *The excursion of a comet* [Hector Servadac, about a trip to a comet]. [...]
>
> That was the sort of science fiction of those days, but in a way perhaps of a more literary nature than the present science fiction. Jules Verne was quite a witty French author. But he is still being read today, my grandchildren enjoy his books as much as I did. [...]
>
> I suppose [I read Flammarion] very soon after I got into reading Jules Verne.

Hector Servadac is not very well known anymore. Oort's grandson Marc Jan Anton Oort, who studied astronomy, told me that in his childhood Oort read Verne's books aloud to him. He inherited a few of these books (but for the two-part Hector Servadac the first part is missing). 'Flammarion' is the book *Astronomie populaire* by the Frenchman Camille Flammarion (1842–1915), a very popular book about astronomy. In Dutch it appeared for the first time in 1884.

> Already in my high school days I got a small telescope from friends of my parents and so I became interested in the stars but not in such a very pronounced way. I liked physics in school just as much and it was perhaps, a more exciting subject matter than astronomy in those days. When I went to the University of Groningen I went there partly because a well-known famous astronomer, J.C. Kapteyn was teaching there. [...]
> Yes, that was the main reason.

From the same diary note of February 6, 1917:

> They want me to study at university: shipbuilding, chemistry, mining engineering, I don't know what. But remember, the elven-rich people's souls are so infinitely more beautiful than shipbuilding, chemistry, the mining industry, remember there's no other way for you than to live a beautiful life in your inner self only: Now then, live.

There is even talk of studying mining engineering in Sweden or Norway and a visit with Oort's father to Professor Knol in Den Haag. Wopke Aurelius Knol (1880–1932) [34] was professor of mining engineering in Delft. Oort summed it up as: 'who did not provide us with a great deal of important information.'

Oort decided to study in Groningen, unlike his older brother Hein who studied physics in Leiden. Although he did not know yet whether he would major in physics or astronomy, he mentioned J.C. Kapteyn's fame as the most important reason to go to Groningen. According to his son Bram Oort, part of the reason also was that he did not want to compete with his brother Hein. In his diary he wrote on September 3, 1917: 'The right choice is to go to Groningen or to go to Utrecht', but without further clarification. More about his choice of university or subject is not to be found in the diaries.

Oort says about his parents in the *AIP-Interview* [6]:

> They were very liberal. They thought that if you really liked the subject very deeply you must follow your calling and try to study that subject. But when at Groningen I wasn't quite decided yet whether I would specialize more in physics or astronomy. It was the personality of Professor Kapteyn which decided me entirely. He was quite an inspiring teacher and especially his elementary astronomy lectures were fascinating.

Fig. 2.0 Oort as a student in Groningen in 1921. From the Oort Archives

2

Kapteyn and Galactic Astronomy Around 1920

> *Those that thoroughly study the lives of those, that*
> *on this Earth did accomplish something significant, are always struck*
> *by the phenomenon that the real strength of these people*
> *has been a shortcoming turned into a quality.*
> Godfried Bomans

> *The wonder is not that the field of the stars is so vast,*
> *but that man has measured it.*
> Anatole France (1844–1924).

2.1 Jacobus Cornelius Kapteyn

Throughout his life, Oort has testified in spoken word and in writing of his great admiration for the scientific approach and work of Kapteyn and his personality, which inspired him to become an astronomer. In order to consider Oort's student time in Groningen, I must first describe Kapteyn's person and his work. There is only room here for a brief treatment, which probably does not do justice to his importance for astronomy at the beginning of the last century and for his significance as the founder of the success of Dutch astronomy in the twentieth century. For more details I refer to my biography of Kapteyn in this series [4] and for a more detailed discussion to the scientific version [3].

Anatole France, born François-Anatole Thibault, was a French poet and novelist. This quote is from *Le Jardin d'Épicure*, translation Alfred Allinson.

© The Editor(s) (if applicable) and The Author(s), under exclusive license
to Springer Nature Switzerland AG 2021
P. C. van der Kruit, *Master of Galactic Astronomy: A Biography of Jan Hendrik Oort*,
Springer Biographies, https://doi.org/10.1007/978-3-030-55548-1_2

A general introduction to the history of astronomy in the Netherlands in the twentieth century has been written by David Baneke [35].

Jacobus Cornelius Kapteyn (1851–1922) (see Fig. 2.1) was born in Barneveld, which is located near the geometric center of the Netherlands. His parents ran a boarding school for boys. After his academic study in Utrecht, completed in 1875 with a dissertation in what we now would call applied mathematics—about the physics of vibrating membranes—, he was appointed as an observator at the Sterrewacht Leiden. In 1878 he was appointed professor at the University of Groningen to teach astronomy and theoretical mechanics. His appointment was a direct consequence of the new law on higher education

Fig. 2.1 Painting of Kapteyn by Jan Pieter Veth (1864–1925). This is a preliminary version of a painting produced in 1918 on the occasion of his fortieth anniversary as a professor, that was presented to his wife. However, Mrs. Kapteyn did not like the painting, and Veth made another one with Kapteyn sitting at his desk (see Fig. 2.8 further on), which she liked much better. In the end an academic gown, jabot and beret were added to the original painting and that is now displayed in the central Academy Building of Groningen University. For the complete story see my biography of Kapteyn [3]. This version is in the possession of Kapteyn's great-grandson, who has been named after him, J.C. (Jack) Kapteyn

(in the spirit of the by then late Thorbecke). The law came into force in 1877. Whereas until then university studies were a preparation for those leading positions in society, for which an academic education was a prerequisite, more emphasis was now placed on scientific research. The aim was shifted from broad general education to scientific training and research. For some time it looked as if the University of Groningen would be closed in that process and only two universities would remain in the Netherlands, in Leiden and Utrecht. The outcome was completely different. Instead, not only was the University of Groningen continued, but the *Athenæum Illustre* in Amsterdam was given the status of a fully-fledged, fourth (albeit municipal) university. Another effect was that the curricula of the national universities were harmonized. As a result of this new law, university budgets and the number of professors increased substantially. Another result was that there would have to be more teaching in astronomy in Groningen and for that purpose a special chair would have to be established.

Kapteyn's appointment in 1878 meant a third professor of astronomy in the Netherlands, but in contrast to those in Leiden and Utrecht, the Groningen professor did not have his own observatory. Kapteyn attempted various times to obtain funding for it. The directors of the observatories in Leiden and Utrecht, however, were reluctant to share the available funds for astronomical research and advised against a third party.

Kapteyn's interest concerned a fairly unexplored field of study of the Universe. Where the vast majority of astronomical research concerned the Sun, the planets and other objects in our own Solar System, Kapteyn was more interested in the distribution of the stars in space. The first things you then need are star catalogues. Now the situation in the eighteenth century was that most work in this area was done in the northern hemisphere, especially in Germany under the leadership of Friedrich Wilhelm August Argelander (1799–1875) of the Bonner Sternwarte. This involved the determination of the exact positions and brightnesses of stars. Argelander's first step was a star catalogue, known as the *Bonner Durchmusterung (BD)*, which contained 324,198 stars.

This was a time consuming, precise and careful job. The position of a star is given by two coordinates, which are called Right Ascension and Declination, comparable to geographic longitude and latitude on Earth (see Appendix A for a more detailed explanation). Each star had to be observed individually with a so-called meridian circle, which was pointing north-south (i.e. on the meridian) and only adjustable in altitude above the horizon. The Right Ascension then followed from a precise timing of the passing of the star through the meridian with respect to accurately calibrated clocks, which indicate the Sidereal Time. In this it is zero o'clock, when the vernal equinox is on the meridian; the vernal equinox is the point in the sky, where the Sun crosses the equator on March 21 or thereabouts, and is also defined as the zero point of Right Ascension. A sidereal day of 24 sidereal hours then lasts on our regular clocks—which run relative to the Sun and not to the stars—for 23 h, 56 min and 4 s. The altitude above the horizon when a star passes through the meridian gives the Declination, which is the angle between the star and the equator on the sky. The Bonner Durchmusterung was the result of many such observations obtained over the period 1849–1863.

But these were stars of course visible from Europe. The work to extend it to the southern skies proceeded very slowly. Soon Kapteyn came into contact with the director of the British Royal Observatory at the Cape in South-Africa, David Gill (1843–1914). The latter had noticed that photographic exposures of the great comet of 1882, which he had made, showed many stars. His idea was to use a special telescope to photograph the entire southern sky, which could then be used to produce a catalogue of the missing parts of the heavens.

But Gill (see Fig. 2.2) was looking for someone to do the work on this. In his letters to Kapteyn he complained all the time about his lack of time and manpower, until finally Kapteyn took the bait and offered that he would do that work. This was the beginning of Kapteyn's Astronomical Laboratory, an 'observatory without a telescope', which was dedicated to measuring plate material obtained elsewhere. The first quotation at the beginning of this chapter refers to this. With funds collected (not only from the Ministry, but from all kinds of foundations), Kapteyn started the project, initially with a single assistant, and after twelve years it was finally finished. In 1900 the third and last Volume of the *Cape Photographic Durchmusterung (CPD)* appeared. Altogether the three volumes contained 454,875 stars. Kapteyn's fame was established.

But he didn't stop there. After all, his real goal was to determine the distribution of the stars in space. But then you have to know distances of stars and that was notoriously difficult. The traditional method to measure distances was using the reflection of the annual motion of the Earth around the Sun, which is an ellipse in the sky. Its semi-major axis is called the trigonometric parallax (see Appendix A) and this becomes smaller when the distance of the star increases. But this could only by applied to a small number of stars. Kapteyn realized that this small displacement of the star in the sky should also have an effect on the time of meridian passage. In principle, you can determine the parallax by measuring that and comparing it with passages from fainter background stars in the vicinity. In the mid-1880s Kapteyn succeeded in applying this method by making use of the Leiden meridian circle, which the director Hendricus Gerardus van de Sande Bakhuyzen (1838–1923) had made available to him during academic vacations (see Fig. 2.3). That was an extremely difficult thing

Fig. 2.2 David Gill, director of the Royal Observatory at Cape of Good Hope. This picture comes from the *Album Amicorum*, presented to H.G. van de Sande Bakhuyzen on the occasion of his retirement as professor of astronomy and director of the Sterrewacht Leiden in 1908 [36]

Fig. 2.3 Meridian circle of the Sterrewacht Leiden (left) and the director Hendricus van de Sande Bakhuyzen at the eyepiece. From the archives of the Sterrewacht Leiden

to do, because even for the nearby stars one has to measure that passage to an accuracy of one hundredth of a second on the clock. This can only be done by repeated measurements, but even then you have to work very consistently. Kapteyn succeeded, so he must have been an exceptionally good observer. His results, however, indicated that the hope of measuring individual distances of stars on a large scale was unrealistic.

Kapteyn therefore took a statistical approach. All stars in the vicinity of the Sun move relative to each other; the Sun at a velocity of about 20 km/s with respect to the average of stars in its vicinity. This is reflected in the small, systematic motions of stars in the sky. These are called the proper motion (see Appendix A) and are measured in arcseconds per year or sometimes to have more suitable numbers per century. If all other stars would stand still, those proper motions would all point in the direction of where the Sun comes from, the so-called Apex of solar motion. Then the magnitude of the proper motion of each star would directly give its distance. Kapteyn made the following comparison. The direct, trigoniometric parallax is based on the radius of the Earth's orbit around the Sun. If you take the motion of the Sun through space, that base is already four times greater per year and furthermore grows with time. But because of the own peculiar motions of stars through space, you can only apply this statistically for a group of stars. This is called secular parallax (see Appendix A for more details) and this is what Kapteyn decided to use.

Kapteyn defined a plan, in which he would use counts of stars as a function of their apparent brightness (in astronomy this is called magnitude, see Appendix A). And in addition he would need large-scale measurement of proper motions of stars. He received a lot of help from Anders Severin Donner (1854–1938) of the Helsingfors Observatory (Helsingfors is Swedish for Helsinki). Donner collected quite a lot of photographic material for Kapteyn, which was then measured in the latter's Astronomical Laboratory.

Kapteyn's plan to determine the stellar distribution in space depended on three fundamental assumptions. The first concerned the distribution of types of stars. These have a whole range of intrinsic luminosities (called absolute magnitude in astronomy). The British, (originally German) astronomer Frederick William Herschel (1738–1822) was the first to systematically make star counts (and catalogues of nebulous objects), which was later extended to the southern hemisphere by his son John Frederick William (1792–1871). Herschel had assumed that all stars were intrinsically equally luminous and that with his telescope he could see to the edge of the Sidereal System. From his counts he found it to be a flattened structure. But when he built a bigger telescope, he saw more stars. Furthermore, he also saw binary stars that rotated around one another in the course of time, and thus should have the same distance. But they could be very different in relative brightness on the sky in spite of having to be at the same distance. Kapteyn now assumed that the distribution of intrinsic brightness or luminosity would be the same everywhere in the Sidereal System. This is called the luminosity function. If one could then determine that distribution locally using secular parallaxes, one would be able to convert counts of fainter stars into a spatial distribution. This assumption has proved to be reasonably correct.

The other two assumptions were problematic. To be able to use the secular parallax statistically, Kapteyn had to assume that everywhere in the system the motions of stars were distributed equally over all directions and on average would be equally large. So isotropic and uniform. And then there was the third assumption, which was that there would be no dust between the stars, that would scatter (or absorb) light and therefore affect the brightness of distant stars. Kapteyn suspected that this interstellar extinction would weaken the light of the stars more in blue light than in the red, so that stars at larger distances would systematically appear redder. He did extensive research on that and found an estimate of the amount of interstellar absorption (as it is usually called, even though it really is mostly scattering), which was not bad at all compared to what we now know.

When Kapteyn looked closely at the motions of stars, he found that they were not random or isotropic at all, but that there were two systematic streams of stars, the directions of which were more or less opposite in the plane of the Milky Way. These are his famous Star Streams. In the end, it turned out that these are not two opposite streams, but result from the fact that the average velocity of stars in the directions towards and away from the center of our Galaxy is greater than in perpendicular directions. Karl Schwarzschild (1873–1916), at that time director of the Sternwarte in Göttingen, quickly proposed this as an alternative explanation. Kapteyn argued that the composition of the two streams in terms of types of stars then had to be the same; this was not the case in the observations available at the time, but it did become clear later that in fact this is true.

Kapteyn's breakthrough came in 1904 when he presented his Star Streams at a major congress in St. Louis, Missouri. This meeting was part of the World Exhibition on the occasion of the centenary (it had been postponed for a year and it was now 101 years) of the 'Louisiana Purchase', where the United States had bought a vast area of land from the French. The organization of the scientific congress was in the hands of Simon Newcomb (1835–1909), director of the Nautical Almanac Office at the United States Naval Observatory (see Fig. 2.4). He and Kapteyn knew each other well from the work of the latter on the *Cape Photographic Durchmusterung*; Newcomb actually had visited Kapteyn

Fig. 2.4 Simon Newcomb, director of the Nautical Almanac Office of the US Naval Observatory and George Hale, director of the Mount Wilson Observatory. These photographs come from the biography Kapteyn's daughter Henriette Hertzsprung–Kapteyn wrote about her father [37].

several times in Groningen. Kapteyn, in one stroke, became a famous and leading astronomer in St. Louis. He was one of the few scientists from the Netherlands (and in fact from Europe) who visited the United States and one of the first in the Netherlands with an Anglo-Saxon inclination, which would more and more replace the orientation towards Europe (in astronomy especially Germany).

St. Louis also had a second consequence of enormous significance. Kapteyn discussed there his *Plan of Selected Areas*. He had selected more than 200 areas of sky, for which he proposed that observatories from all over the world would determine brightnesses and colors of stars to as faint as possible, spectral types (what the spectrum looked like), measure parallaxes (if possible), proper motions and radial velocities. In the meantime it had been possible to record spectra of stars on a large scale. In these dark absorption lines are present which come from the outer layers of the star, absorbing light from the stellar surface. These provided much information about the nature of a star. But one can also compare the exact wavelength of these lines with those in the laboratory. This allows the measurement of the radial velocity of the star—towards or away from us along the line of sight (see Appendix A). Kapteyn's laboratory would again take care of a large part of the measurements on the photographic plates. All this information could eventually be used for a comprehensive analysis of the structure of the Galaxy.

Kapteyn managed to get his *Plan of Selected Areas* adopted by a large number of important observatories. Edward Pickering (1846–1919), director of the prominent Harvard College Observatory, offered his cooperation. With his observatory and its station on the southern hemisphere (at that time in Peru) photographic plates were taken of all *Selected Areas*, which would eventually (most of them after Kapteyn's death) give uniform coverage in star counts across the entire sky. Of spectacular importance was that Kapteyn met George Ellery Hale (1868–1938) in St. Louis (see Fig. 2.4). He was director of the Mount Wilson Observatory near Los Angeles, California, where he was building at that time the world's largest telescope, the 60 in. Telescope. Hale adopted Kapteyn's *Plan of Selected Areas* as the primary observing program for this new, giant telescope and offered Kapteyn a part-time position as 'Research Associate' of the Carnegie Institution of Washington (that financed of the Mount Wilson Observatory). This made Kapteyn one of the most prominent astronomers in the world. Now the counts could be done even deeper then at Harvard (at least as far as visible from Mount Wilson). From 1908 onward, Kapteyn went to

Fig. 2.5 Jacobus Kapteyn and his wife on Mount Wilson during their visit in 1909, in front of the tent that had been put up for them. From the University Museum Groningen

Mount Wilson every year for several months. From the second time onward (in 1909) his wife accompanied him; first they camped in a tent on top of the mountain (Fig. 2.5), but the next time a special small house, the Kapteyn Cottage, had been built for them. After 1914, the First World War made it impossible for Kapteyn and his wife to cross the ocean.

2.2   The Kapteyn System

Kapteyn's *Plan of Selected Areas* was going to be a huge job and there was no chance that sufficient progress would have been made to draw conclusions when Kapteyn were to retire in 1921. But there was more work going on as well, although not as systematic as the Plan. For years, Kapteyn had been working with photographic plates from elsewhere, measuring in particular proper motions and thus statistical distances of stars using his method of secular parallax. Kapteyn's student Herman Albertus Weersma (1877–1961) obtained his PhD under Kapteyn in 1908 on determinations of the motion of the Sun through space, information which is essential for the application of the method of secular parallax. Weersma worked for Kapteyn in Groningen until 1912, when he left astronomy. Kapteyn's student Pieter van Rhijn obtained his PhD in 1915 on a thesis consisting of two parts. The first part dealt with the issue of absorption of starlight determined by the color of stars as their distance increased. He actually confirmed what Kapteyn had already found, its existence as well as the amount of it per unit distance. The second part was about a necessary preparation for a study of the distribution of stars in space, namely the determination of the average distance (or parallax) of stars as a function of their apparent magnitude and proper motion. If one were to take stars of a certain apparent brightness, what is the average distance? Or for stars of a certain proper motion, what is the average distance? With a reasonable assumption about the intrinsic spread of these properties among stars around their averages, you can determine the brightness function, which is the distribution of intrinsic luminosity of stars. The work of Weersma, van Rhijn and Kapteyn culminated in a number of articles between 1916 and 1920, which made a first analysis possible (Fig. 2.6).

That analysis was done in two steps. Kapteyn and van Rhijn did the first part together. Using all the preparatory work, they first determined the luminosity function. It turned out this could be very well approximated by a simple Gauss-curve, the known distribution of odds in a random process. The brightest stars are very rare, but numbers are increasing for fainter stars. This then reaches a maximum for stars intrinsically twenty times fainter than the Sun and decreases

Fig. 2.6 Herman Albertus Weersma and Pieter Johannes van Rhijn, Kapteyn's assistants during 1908–1912 and 1914–1921 respectively (for the period 1912–1914 Frits Zernike held the position). Weersma: from the *Album Amicorum*, presented to H.G. van de Sande Bakhuyzen in 1908 [36]. Van Rhijn: from Wikimedia Commons [38]

again for even fainter stars. They then used the star counts and averaged these for a strip along the Milky Way. They repeated this for strips around Galactic latitudes (height above or below the plane of the Milky Way on the sky) of $\pm 30°$ and $\pm 60°$ and for an area around the poles at $\pm 90°$. And from this they determined for each of these four areas the density of the stars as a function of the distance from the Sun. The result was then the distribution of the stars in the system as shown in the upper panel of Fig. 2.7, which they then published in 1921.

There is the following to note about these calculations. In the first place, Kapteyn and van Rhijn averaged over the entire Milky Way and parallel to it, and above and below. So the model automatically became symmetrical. Together with the calculation of density as a function of distance to the Sun, this immediately implies that the Sun is in the center. Finally, they neglected the absorption of starlight. Why did they do that? After all, studies by both of them had provided strong evidence for its presence and had come up with more or less the same amount per unit distance. Well, there had been work by the American astronomer Harlow Shapley (1885–1972), then of the Mount Wilson Observatory, who had made a study of globular clusters. These are clusters of about one hundred thousand stars or sometimes more with a spherical structure, usually lying outside the Milky Way. In Messier 13 in the constel-

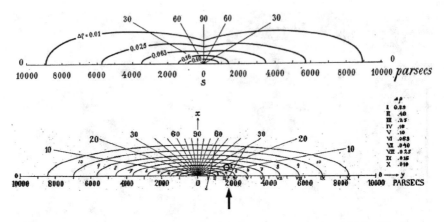

Fig. 2.7 The Kapteyn system. Above the model from star counts as published by Kapteyn & van Rhijn in 1921, and under after fitting of ellipsoids by Kapteyn in 1922. The circle near the arrow in the lower panel shows the position of the Sun, but is drawn three times too far from the center. Kapteyn Astronomical Institute, University of Groningen

lation Hercules he determined the colors of 1300 stars. Now the distance was determined using variable stars in it (RR Lyrae stars), which pulsate and of which the period of the variation is a measure for the absolute magnitude of the star. The distance could thus determined to be 10,000 parsec (or 10 kpc; a parsec is the distance of a star with a parallax of 1 arcsec, equivalent to 3.26 lightyears, see Appendix A). This is now determined to be 6.8 kiloparsec (kpc). Shapley found that if the absorption, as Kapteyn had found it to be, were to be applied, even the reddest stars in the cluster would intrinsically have to be much, bluer than the bluest stars we see in the solar neighborhood. This meant that the interstellar absorption had to be hundreds of times less than Kapteyn had determined. This was convincing enough for all astronomers, including Kapteyn, to neglect absorption. Kapteyn had always said that the color change he had found on average with distance could also be the result of a systematic change in the type of stars at a larger distances. The dust is limited to a thin layer in the Milky Way and is indeed very important in the plane of the Galaxy, but negligible in directions below or above it.

In 1922, the year of his death, Kapteyn (Fig. 2.8) published an article in which he studied the dynamic structure for the first time. This can be seen as the first application of stellar dynamics to observational data. He did this as follows. He now knew the spatial distribution of the stars. He first looked at the vertical structure in his model for the Sidereal System and calculated the gravitational field that would correspond to it. In order to do so, he approached the star distribution with ellipsoids (Fig. 2.7, lower panel). This gave the vertical force except for a factor, which corresponded to the average mass of a star. If

Fig. 2.8 Kapteyn behind his desk. This painting by Jan Pieter Veth is displayed in the Kapteyn Room of the Kapteyn Astronomical Institute in Groningen. The person in the top right corner is David Gill. Kapteyn Astronomical Institute, University of Groningen

everything was in equilibrium, the distribution of the stars would have to be in agreement with the distribution of the velocities. If the average velocity is larger and the gravitational field the same, then stars will be able to attain greater heights before they fall back to the plane, so the vertical distribution will be thicker. The average velocity was known from measurements of velocities of stars and Kapteyn found that the average mass of a star had to be 1.4–2.2 times that of the Sun to reach a situation of equilibrium. That seemed reasonable to him because from orbits of binary stars it had been found that together they have on average a mass 1.6 times that of the Sun. Then most of the stars should actually be a binary star.

Kapteyn used this subsequently to calculate the force in the direction of the plane and then found that for equilibrium it was necessary that there should be an additional centrifugal force corresponding to a rotation velocity of about 20 km/s. For various reasons he deduced that the Sun could not be exactly in the center of the system, but about 600–700 pc away from it and 40 pc outside the plane. Now Kapteyn's Star Streams had a relative velocity of 40 km/s, while the centrifugal force required from equilibrium was calculated to be about half of that. Kapteyn therefore deduced that stars in the system near the Sun rotated about the center at 20 km/s, one half in one direction and the other half in the opposite one. In other words, he found a consistent model. The direction towards the center then had to be perpendicular to the Star Streams, which left two choices; Kapteyn inferred it had to be the direction of the constellation Carina in the southern hemisphere, where the Milky Way is relatively bright.

The British James Hopwood Jeans (1877–1946) is one of the founders of stellar dynamics (now usual referred to as galactic dynamics), which describes star clusters, galaxies and clusters of stars as an equilibrium between motion and gravity. From him comes the term Kapteyn Universe. Kapteyn's early work on his the distribution of stars and motions, and Jeans' detailed treatment in an important paper in 1922 (entitled *Motions of stars in a Kapteyn Universe* [39]) was the crowning achievement of Kapteyn's work. It must be said that the German astronomer Hugo Hans Ritter von Seeliger (1849–1924) at about the same time derived a comparable, but more schematic and mathematical model. It had a less direct relation to the observations and quickly became forgotten.

Stellar dynamics appealed very much to Oort. In his 1922 article, he suggested that the sudden transition at about 65 km/s of an isotropic (equally distributed over all directions) distributions for lower-velocity stars to one limited to only one hemisphere for higher velocities was due to the fact that the 65 km/s or so would correspond to the escape velocity from the Kapteyn system. The high-velocity stars would then be stars that did not belong to the system (he called them 'strangers'), but were part of a much larger structure. The limitation in the directions would then be the result of a systematic velocity of the Kapteyn system with respect to this larger system. Kapteyn had calculated that for *his* system the escape velocity would be 81 km/s (actually, that value had not been published, but he will have told Oort) and Oort thus deduced that the average stellar mass had to be 0.65 times that of the Sun.

All in all, the picture was that of a Kapteyn System that was flattened and dynamically stable. In the vertical direction, the velocities of the stars were such that their spatial distribution and their gravitation were precisely in balance. In the direction *in* the Milky Way plane a centrifugal force was needed, which resulted from rotation around the center in two opposite directions.

3

Student at the University of Groningen

It is the supreme art of the teacher
to awaken joy in creative expression and knowledge.
Albert Einstein (1879–1955)

Talent alone won't make you a success.
Neither will being in the right place at the right time,
unless you are ready.
The most important question is: 'Are your ready?'
John William (Johnny) Carson (1925–2005)

3.1 Student Life in Groningen

Oort began his studies at the University of Groningen in September 1917. Kapteyn then still had four years to go before his retirement. At first Oort lived at the address Poststraat 2; this is on the corner with the Oude Boteringestraat, not much more than a block from the Academy Building and the Astronomical Laboratory. In a letter to his brother Hein he wrote about the accommodation; the room had a chimney, a table in the middle, a sofa, a piano, and so on. Apparently there was a separate bedroom and washing facilities, and it was quite a spacious room. According to the 'Groningsche Studenten Almanak' he did not live there long; after a short period at the Martinikerkhof 17a, he lived the rest of his Groningen period at the address Brugstraat 7a. This is close to the center of town.

His letters to his parents and to Hein (there are letters from his brother John as well, but these are mainly about biological subjects) show that he led an

© The Editor(s) (if applicable) and The Author(s), under exclusive license
to Springer Nature Switzerland AG 2021
P. C. van der Kruit, *Master of Galactic Astronomy: A Biography of Jan Hendrik Oort*,
Springer Biographies, https://doi.org/10.1007/978-3-030-55548-1_3

active social life. Besides going to concerts he was also a member of a dance club. On 22 October 1918 he wrote to Hein:

> Furthermore, I often play the piano in the morning and I am especially concentrating on the finger exercises. That sonata seems difficult to me, but I will do my best. We have a nice dance club, so now I get to know more Gron. girls and there are very nice ones here. But since we've only have had one dance lesson and I didn't know any of them before, I'll be able to tell you more only later on.
> Do you have a lot of time for reading? I do not have very much time, or rather I am not getting around to it at all.
> In the Harmonie I recently [heard] a very nice suite by Bach, for timpani, etc. You know, the suite that contains the violin aria that you played so often.

The Bach Suite must have been BWV 1068 in D major and the violin aria the 'Air' in it.

Oort was a member of the Groninger Student Association *Vindicat atque Polit* (Maintain and Civilize), but whether he spent much time at the society *Mutua Fides* (Mutual Trust) is not clear from the letters. In the photograph in Fig. 4.0 between Chaps. 3 and 4 we see Oort wearing a 'corps hat'. Anyone who applied for membership of the fraternity Vindicat at the time, had to wear a flat cap, but when the initialization with the 'actus' as its final act had passed and they had become full members, the cap was exchanged for a corps hat, which every 'Vindicater' would then wear at official ceremonies (e.g. parade at the annual dies natalis, the anniversary of the University). During the First World War, such official ceremonies were suspended. This is probably why the picture was taken inside, probably in Mutua Fides' conversation room. Oort certainly participated in the most important events, such as the annual gala, a ball with the participants in gala dress (see Fig. 3.1). Because Vindicat was only for men the members invited a lady to accompany them, in many cases recruited from the members of the Groningen Female Student Club *Magna Pete* (Aiming for the Great). Usually it was preceded by a concert in the concert hall of the *Harmonie* with a dinner afterwards.

It was not cheap to have your children pursue academic studies. In the Oort family, all three boys went to university (Hein physics in Leiden, Jan astronomy in Groningen, and John biology in Utrecht). At the time, this cost something like 1200 guilders (comparable to 6800 € today) per year for each of the boys. Students often had an additional income (tutoring, mentoring), but it seems that Oort did not. Father Oort's salary as director of Rhijngeest was certainly not low, but this must have been quite an expense. The two girls did not go to university (they worked in social work or health care).

Father Oort wrote on 8 April 1919 in a typed letter:

Fig. 3.1 Part of a photograph taken during a gala of the Groningen Student Association *Vindicat atque Polit* in 1920. From the Oort Archives

Dear Jan, First of all, I would like to congratulate you on your successful examination; the number of these things will grow if you do them separately for the various courses. At least I do not want to give you some extra money as appreciation for this on a case-by-case basis. Yet I hereby send you f 80.00. In the hope that this will be sufficient to meet the necessary needs. Mother advised me to put the pen on your nose as well, that you have to learn to match your expenses with your income, because otherwise you will have to come and study in Oegstgeest, and so on. But on the other hand, as you can understand, mother was the first to say: send him something as soon as possible and noted as positive point that it is so good that you do not get into debts. She later found out that among what you had have paid, there were bills from last year, but anyhow. I will not elaborate any further: you will understand and at least we both are aware that you should not be ranked among the wasters. Furthermore, Mother is anxious for you to send your laundry before Saturday: she waited in vain for it this week.

In an undated letter (which must have been written in the fall of 1918, because it refers to the end of the War):

How wonderful life is!

It's as if music has the power to reveal all the deepest secrets and connect all feelings. There is something great about it that makes you live and gives you character! That when you do the wrong thing, you are deeply sorry at the same time. […]

When the armistice was announced, the Martini Tower started ringing, in the middle of breakfast; I had never heard its bells ring, it was a beautiful sound, very deep and low. They seem not to have been used since they announced the start of the war in 1914. It felt more like sadness than about celebrating. But all the schools closed and the people of Groningen seemed to think differently. I never saw so many flags.

This is by the way the only reference to the First World War in the Oort's diaries! The impact on daily life has apparently not been that great for high school and university students.

Oort was active in the Groningen Student Rowing Club *Aegir*. In Norwegian mythology, *Aegir*, which means 'sea', is a giant living in the ocean. The club was founded in 1878 under the authority of *Vindicat*. It also became active in rowing competitions. The most important one is the Dutch 'Varsity' race, a rowing regatta (series of races) between students of various Dutch universities. In Oort's time it was held on the North Sea Canal, that connects Amsterdam to the coast at IJmuiden, in the second half of May or early June. It had grown from a challenge in 1878 between *Njord* and *Laga*, the equivalents of *Aegir* in Leiden and Delft respectively. From 1882 onward, the Utrecht club *Triton* also participated, but for *Aegir* it was too far from Groningen to participate according to the Groningsche Studenten Almanak; this probably refers to the cost of transporting the boats. The number of classes of boats also increased and formally the Vasity resulted from this, for the first time as such in 1883.

The most important is the 'Old Four', which means four rowers with one oar each. The coxswain (or cox for short) has a rudder and sits in the stern. 'Old' refers to the best and most experienced team. The distance covered was three km. *Aegir* participated in the Varsity for the first time in 1918 and Oort was the cox of the Old Four. *Nereus* (Amsterdam) was also a participant in the meantime. *Aegir* won the race with six boat lengths and thus also the Varsity (Fig. 3.2). It would take until 1956 before they would win again.

A year later Oort participated again, but then they finished third. He was also the cox of the 'Young Four' at the time, but they didn't win either. In 1920 the Old Four did not participate because of illness; for 1921 the Groningsche Studenten Almanak only mentioned that the *Aegir* team had not done so well. It turns out that in the skiff the honor of *Aegir* was saved by one E.A. Kreiken. This was Egbert Adriaan Kreiken (1896–1964), also an astronomy student. Interestingly, Kreiken had been born in the same house in Barneveld as Kapteyn (see my Kapteyn biography [3] for details). Kreiken was a few years

Fig. 3.2 The 'Old Four', with Oort as cox, of the Groningen rowing club *Aegir* after they have won the Dutch Varsity edition 1918. From the Oort Archives

more senior than Oort; he obtained his PhD at the University of Groningen in 1923.

And there was skating, another great love of Oort. I quote from a letter to his parents dated February 5, 1922:

> Today two weeks ago I skated again on the Paterswoldse Meer [a large lake to the south of Groningen] with Schilt, as I wrote to you. The following week that followed was so terribly cold, that every step in the street was a misery. [...] Piet and I took a skating trip anyway on Wednesday across the Reitdiep, sheltered by high dikes. The ice was incredibly thick, but bad with floes and also holes in the ice because of the strong wind. There were no other skaters and we got a lonely and deserted feeling, in that weather! Friday the wind dropped a bit and the temperature rose and I could not resist the temptation to skate down the Damsterdiep, where there was very good ice, only a lot of wind. Coming from the ice, Piet and I suddenly felt like dancing and – we could not be more fortunate – the same evening there was a masquerade ball in Frigge for members of the Clubhouse, where by the way we had never planned to go. We called the chairman and we could still obtain tickets, the enthusiasm spread to four other housemates and we went there. We rented dominoes there, everything very easily arranged and then dancing. It was extraordinarily nice and in hindsight I rarely had such a great ball. Costumes were nice and the crowd was very nice,

many acquaintances of course. After the demasqué, it got even more exciting. I dined with Annie Mossel, a beautiful girl but engaged, and Piet with Toet Mees. August Maes, one of our former housemates, invited us to his table, where there were a few girls too many. The dinner was very animated, except that we ate little, since serpentines were thrown, so that all the tables were covered under a meters long layer of paper, and a lot of dancing was done.

The dancing went on until 4 a.m.and yet I was at the lab at 9:00 a.m. on Saturday morning.

Apparently, Piet lived at the same address as Oort. It is probably Pieter Hendrik de Waard (1899–1980), student of law. Jan Schilt, another more senior astronomy student, we will meet again regularly. A domino is a costume for a masked ball, including a mask.

3.2   Kapteyn's Lectures

In the Oort Archives there are notes of the lectures of Kapteyn that Oort attended. There are two notebooks, one of which has been started from both sides. These contain two Kapteyn lecture courses: 'Popular Astronomy' is divided among two notebooks and 'Spherical Astronomy'. In the last course, Kapteyn discussed everything related to positional measurements in astronomy, such as spherical trigonometry, astronomical coordinates, etc. The lecture course on Popular Astronomy, which today would be called General Astronomy, dealt exclusively with the planetary system. Kapteyn treated the history of astronomy up to the then current knowledge of the Solar System and its objects. It must have been these lectures that made such an impression on Oort that he decided to take astronomy as his main subject. I quote from an autobiographical piece written by Oort in 1981 [40]:

> When in 1917, at the age of 17, I began my studies in Groningen I became almost immediately inspired by Kapteyn's lectures on elementary astronomy. Although I had been strongly interested in astronomy since my high school years in Leiden, and this had influenced my choice of the University of Groningen because Kapteyn was there, I was, in 1917, still undecided between physics and astronomy as my major direction; I remember that I was so impressed by the way he taught elementary celestial mechanics that I tried to convey my new insight to friends who had likewise just entered the University, but were studying humanities. But I do not believe that I succeeded in conferring to them the full appreciation of the fascination of celestial dynamics.
>
> Perhaps the most significant thing I learned – mainly, I believe from Kapteyn's discussion of Kepler's method of studying nature – was to tie interpretations directly to observations, and to be extremely wary of hypotheses and speculations.

In the first part of his course, Kapteyn refrained, for instance, from introducing the notion of 'force' to replace the measurable quantity 'acceleration'. He disliked intricate mathematical formulations which prevented one from 'seeing through' a theory; he feared the danger that the formulae might make one lose sight of the essentials. This was, of course, before quantum mechanics brought home the fact that one's sight is insufficiently developed to 'look through' the deeper domains of physical science without the aid of mathematics.

What did Oort refer to with Kepler's method? In his lectures Kapteyn dealt with the changing insight into the structure of the system of Sun, Moon and planets. The work of the classical Greek philosophers had resulted in the view summed up by Claudius Ptolemæus (Ptolemy) (c.100–c.170) in his famous book 'Almagest', in which the Earth was at the center and the Moon, Sun and planets orbited around it. For example, the outer planets, of which we now know that they are further away from the Sun than the Earth, described orbits that were a superposition of a large circle, called the deferent, and a smaller one, called the epicycle. The deferent is actually the orbit of the planet itself and the epicycle is a reflection of that of the Earth, from which we observe. If you don't know more than positions in the sky (i.e. directions and no distances), that is in principle an excellent description. It was originally assumed that those motions of celestial bodies would not only take place in perfect circles, but also at precisely uniform speeds. To allow for the fact that in reality planets do not move in such manners, Ptolemæus (Ptolemy) had adapted this view by removing the Earth from the center and making the motion uniform with respect to a point called 'equant' opposite the Earth. Nicolaus Copernicus (1473–1543) did put the Sun in the center, but held on to the uniform circular motion. The accurate observations of the positions of planets obtained by Tycho Brahe (1546–1601) painfully revealed the inaccuracies.

Johannes Kepler (1571–1630) used Brahe's observations of Mars and found—as Oort wrote down in his lecture notes (his underlining) –:

> Kepler now became convinced that something had to be wrong with this system, because the observations never completely corresponded with the theory. Kepler then started all over again, accepting as little as possible, and determined the orbits <u>from</u> his <u>observations</u>, without accepting any assumptions.
> What Kepler did assume were two things:
> 1^e that the orbit of Mars is flat.
> 2^e that it has a closed orbit.

You cannot proceed completely without assumptions, but you can keep them to a minimum, accepting only the absolutely necessary ones. For centuries it had simply been assumed that the ideas of uniform circular movements in the orbits for objects beyond the Moon (the superlunal) had to be correct, going back to classical philosophers like Plato (427–347BC) and Aristotle (384–322BC). Kepler was the first to ignore such preconceived ideas and based his work solely on observations to determine the shape of the orbit. This is what Oort meant; that, as he often emphasized, Kapteyn taught him to stay as close as possible to the observations.

Oort passed this on to his students. As a student I attended his lecture course 'Planetary System' in Leiden in the academic year 1964–65. My notes of Oort's lectures and the notes of Oort of the lecture by Kapteyn show a significant degree of similarity. Like Kapteyn, Oort showed in detail how Kepler proceeded, and how he determined the orbit of Mars from Brahe's accurate observations and in this way found his first two laws of planetary orbits. The method used by Kepler is explained in detail in Appendix A.1. Kepler's first law says, that the orbits are ellipses with the Sun in one of the foci, and the second, that the orbital velocity varies with distance from the Sun, such that it is inversely proportional to the distance and thus that it is greatest when the planet is closest to the Sun. The latter is now known as the conservation of angular momentum.

Oort's admiration for Kepler's work, as conveyed by Kapteyn, was thus conveyed to his students, at least to me. Later I also read this history in the excellent and fascinating book *The sleepwalkers: A history of man's changing vision of the Universe*, written in 1959 by Arthur Koestler (1905–1983), which was started as a biography of Kepler [41]. I have given copies of that book as a present to all my PhD students after they had defended their dissertation. My fascination goes back to the lectures of Kapteyn, transferred to Oort and through him to me.

Kapteyn had thus made a lasting impression on Oort and the latter has often emphasized this in his presentations and writings. On the other hand, Kapteyn was impressed by Oort and, as we shall see, he strongly recommended Oort for a job at the Sterrewacht Leiden. Oort's first publication was about Kapteyn. It is an article in the *Groningsche Studenten Almanak* for the year 1922, on the occasion of Kapteyn's retirement at the end of the academic year 1920–1921, in which Kapteyn had turned seventy [42]. This article starts with a theme that is often found in Oort's lectures and publications, even in his lecture in 1988 at the presentation of the Kyoto Prize, which was entitled *Horizons*.

Many of you, looking out over the sea, have sometimes felt a desire to discover the unknown, that which is hidden behind the horizon.

A large overview of the whole and insight into what can lead to great discoveries in the future, are the main features of Kapteyn's work. This clarity of insight characterizes him. Every problem, no matter how complicated, is directly linked

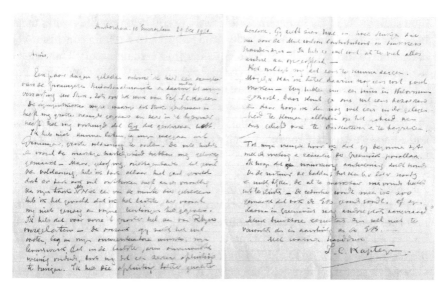

Fig. 3.3 The only letter from Kapteyn to Oort. The content is reproduced in the text, translated into English. From the Oort Archives

to the basic problems, stripped of the side-effects that deprive others of seeing a way towards the solution.

Kapteyn responded with the only known letter between them (see Fig. 3.3). Because of the special relationship between Kapteyn and Oort, I include the text here in full in English translation. Kapteyn was by then terminally ill and was nursed in his daughter's home in Amsterdam. He never lived in the house he and his wife bought in Hilversum.

Amsterdam. 10 Emmaplein 29 Dec 1921

Amice,

A few days ago I received a copy of the Groningsche Studenten Almanak and to my surprise found your article, Something about the work of Prof. J.C. Kapteyn. The sympathetic manner in which the piece has been written has given me great joy and, in particular, it has pleased me that it was You that wrote it. When I left Groningen, I could not help but feel a great sense of satisfaction. The many tributes and especially the great cordiality have made me happy. But, believe me, notwithstanding the feeling of satisfaction, I also realize very well that some things has been lacking in the fulfillment of my task. Not one of the least flaws I have felt was that I have the last year not given enough attention to my students. I left that for by far the greatest part to van Rhijn. – The cause, thou shalt know it, lay in my irresistible wish, to still bring my life's work, that in recent years had made unexpectedly little progress, to a certain conclusion. I have come to such a conclusion taliter qualiter [to a certain extent]. Thou shalt see how in two pieces that are now in print for the Mount Wilson Contributions – I have neglected a little too much of everything else for that. –

It is good for me to have a chance to say this. Maybe I can make up for it later. – We have bought a house in Hilversum. Maybe you can visit us there one time, and I hope we can have the opportunity to discuss all sorts of things about our beloved profession.

———

I am pleased to hear that you have started measuring and reducing the Greenwich parallaxes. I hope that you will keep a close eye on the nature of the images; e.g. it may later prove necessary and useful to exclude the all too noticeably deformed images – I think the reduction is done in such a way that also the EB [PM, proper motions] can be found. Or have any other plates been requested from Greenwich for that purpose? Completely usable results may not be expected other than in connection with the EB.

With a warm handshake,

J.C. Kapteyn

Pieter Johannes van Rhijn (1886–1960) became Kapteyn's successor as professor of astronomy and director of the Astronomical Laboratory (see Fig. 3.4). During his long period as director (1921 to 1957), the Groningen Institute lost its prominent position in the international arena, which was taken over by the Sterrewacht Leiden.

3.3   Astronomical Research

Oort's studies went along successfully. In the *AIP-Interview* [6] Oort was also asked about his contacts with other professors (in addition to Kapteyn and van Rhijn), while a student in Groningen.

> Yes. I was interested in mathematics. I even went to lectures given by a famous French Professor Denjoy in Utrecht on point ensembles. But that was an interest of a more aesthetic nature and not an interest that would ever lead me to take up mathematics as a profession. I was too interested in nature itself. There was one professor in Groningen who certainly had considerable influence on my education. This was Professor Zernike, a theoretical physicist and also experimental physicist. He received the Nobel Prize for the phase-contrast microscope that he invented, but is nowadays best known for his theoretical work in optics. In a way he was the worst teacher imaginable, he never prepared his lectures

and often repeated the same stuff over again the next week because he had not prepared anything new. Many people found him difficult to follow in lectures but I learned more from him than from almost anybody else.

Oort's curriculum included—in addition to Kapteyn's lectures—mathematics lectures and practical instructions from Professors Johan Antony Barrau (1873–1953) and Julius Wolff (1882–1945), physics from Professors Hermanus Haga (1852–1936) and Frits Zernike (1888–1966) (see Fig. 3.5). Chemistry was taught by Professors Frans Maurits Jeager (1877–1945) and Hilmar Johannes Backer (1882–1959). The curriculum also included some mineralogy and geology, and the professor was Jan Haitzes Bonnema (1864–1941).

Oort passed his candidate exam (equivalent to bachelor) in February 1919 (*cum laude*). In that academic year, there were a total of only three 'cum laudes' out of a total of 62 candidate exams in Oort's Faculty (Mathematics and Natural Sciences). Apparently there were no lecture courses in astronomy for the doctoral exam (equivalent to master), which is perhaps not surprising because of the small number of students. He came to work daily at the Astronomical Laboratory (Fig. 3.4), where he was involved in the research program. He conducted his first research with senior student Jan Schilt (1894–1982), see

Fig. 3.4 The Astronomical Laboratory, after it was named after Kapteyn in 1921 on the occasion of his retirement. It was located diagonally behind the Academy building in the Broerstraat and served as a home for the Groningen astronomers until 1968, when they moved to the campus in the north of the city. The university seized the opportunity to demolish it after a fire on the upper floor in 1988. University Museum Groningen

Fig. 3.5 Hermanus Haga and Frits Zernike, whose physics courses Oort attended. From University Museum Groningen

Fig. 3.6. Schilt went to work in Leiden in 1922, from where he obtained his PhD in Groningen with van Rhijn as a supervisor. He spent the rest of his life working in the United States, eventually as director of the Rutherford Observatory at Columbia University in New York.

The research Schilt and Oort did was an entry for an essay prize, issued in 1921 by the Bachiene Foundation of Leiden University. This fund was set up in the testament of the politician and Thorbecke supporter Philippus Johannes Bachiene (1814–1881). His daughter, Marianna Carolina Francisca Baart de la Faille-Bachiene (1847–1912) had earmarked the prize for research in human physiology, zoology or astronomy for students who could use financial support. The subject of the prize for 1922 had been defined by Kapteyn (and published in a British journal, so it was open to foreign entries) and included a discussion of the spatial velocities of stars. Schilt and Oort used an existing catalogue with motions in the sky (proper motions) of more than 6000 stars, which was compiled and published in 1910 by the American astronomer Lewis Boss (1846–1912), director of the Dudley Observatory in Albany, New York. The prize, which they indeed won, was a certificate, which would have gone to Schilt, and a sum of five hundred guilders (now corresponding to about 1750 €). What happened to the latter is not recorded. The work was printed as a separate publication.

It is not clear how many contestants there were. The current treasurer and chairman of the foundation, nor the Leiden University Library, were able to tell me where the archives of the foundation are. There is a remarkable story of a contemporary of Schilt and Oort, Willem Jacob Luyten (1899–1995).

Fig. 3.6 Jan Schilt (left) and Willem Jacob Luyten. Schilt's picture comes from the group photograph of the General Assembly of the International Astronomical Union at Harvard University in 1932. Luyten's is from 1928. Both are from the Oort Archives

He (see Fig. 3.6) wrote in his memoirs (privately published and titled with some self-mockery, *My first 72 years of astronomical research: Reminiscences of an astronomical curmudgeon, revealing the presence of human nature in science* [43]):

> During my last year in Leiden, Kapteyn retired from his professorship at Groningen and came to the Observatory at Leiden as a consultant. That year, the Bachiene Foundation, which annually issued a call for a prize essay in one of the sciences, announced one in astronomy. The Foundation was asking for a thorough discussion of solar motion and individual stellar velocities of the stars in the Boss Preliminary General Catalogue.
>
> As my interest in stellar motions had awakened, I thought of writing in for it and asked Kapteyn what he thought of the idea. He answered, 'You are perfectly free to do so, but I should tell you that at Groningen we have one copy of the catalog, in which all the Upsilon and Tau Components which we calculated have been entered, and we have given this copy to Schilt and Oort. Since these Upsilon and Tau Components were absolutely essential in writing the prize essay and it would have taken me hundreds of hours to recalculate them, it was plain what Kapteyn meant.
>
> This was my first introduction to high level scientific manipulation. I never forgot it. In due course, Schilt and Oort got the prize.
>
> Many years later a similar essay prize was called for on the variable star RZ Cassiopeia. The only person who had all the available observations was Aernout

de Sitter, son of the Director of the Observatory, You can guess: Aernout de Sitter got the prize.

Aernout de Sitter (1905–1944) and Oort would later become very close friends. Luyten was a special character. When he visited Harvard, where Luyten worked at the time, Oort described him in his pocket diary for 1924 as follows:

> Continuation Friday, March 21. [...]
> I talk to Luyten a lot, although he is always looking for an argument with anyone and always wants to talk about the lesser qualities of others. In addition, he often boasts a little and is not very civilized in his choice of words. But he is pleasant company and has a good opinion on many things and people, and also has a good taste for music and many other things.

The first scientific article, which Oort wrote by himself, was published in the first volume of the *Bulletin of the Astronomical Institutes of the Netherlands (BAN)*, and is dated September 1922. The subject was the stars with exceptionally high velocities. When asked when he first did research under Kapteyn, Oort replied in the *AIP-Interview* [6]:

> Early in my third year, I thought. [So late 1919.] The first two or three years usually are taken up by a kind of general physical mathematical education at universities and it is only after the first examination, Candidates examination, that research is taken up by the students. [...] My own earliest research was on the phenomenon of stars of high velocity which certainly fell outside the regular domain I would say of research of the Laboratory at that time. But it was certainly close to Kapteyn's interests.
> Yes, I knew fairly well what they were doing at the time at the laboratory and had some discussions with him [Kapteyn] on this subject, but only at the end when it was finished and he could give an opinion on the possible interpretation which was valuable to me at the time. But I certainly don't want to imply that there was any difficulty in approaching him. It was just perhaps, ordinary shyness of a young student. [...]
> The stars of high velocity [...] turned out to be a subject of rather fundamental importance which I didn't foresee when I started to work on them at that time. I was just puzzled by it. But in essence they foreshadowed the rotation of the Galaxy which at that time had not yet been conceived.

For the time being, the high-velocity stars remained a mystery. Oort's most important finding is shown in Fig. 3.7. What he had done was take collections of stars, whose spatial velocities were known with reasonable certainty. This means that three things had to be known. The first is the proper motion; this is the displacement over time of a star in the sky as a result of its motion in space. If that is known (in arcseconds per year, for example) and if, secondly, the distance is known, one can calculate the velocity in km/s perpendicular to

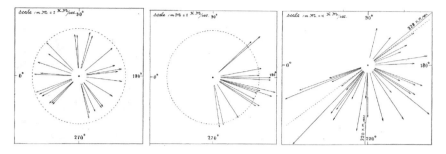

Fig. 3.7 Figures from Oort's 1922 publication on high-velocity stars. See text for further explanation. From the Oort Archives

the line of sight. This is called the tangential velocity. Thirdly, one needs to know the radial velocity, i.e. along the line of sight away from us or towards us. This is measured with the Doppler effect, which causes the wavelengths of the lines in the spectrum of a star to change. If you have all that information, you know the size and direction of the velocity of a star in space (relative to the Sun, but because the velocity of the Sun was known, you can correct for that and find it relative to the average of the nearby stars).

Oort collected all this information and what he then did was calculate the direction of the motion of the star projected on the plane of the Milky Way and seen from the Sun. Figure 3.7 shows this for a group of stars, which are well-selected so that the set is statistically complete for an peculiar motion greater than a certain value. This figure is a bit strange, because the orientation of each arrow indicates the direction of the *projection* of the velocity on the plane of the Milky Way, but the length indicates the *total* spatial velocity. On the left, only stars have been plotted that have a spatial velocity less than 62 km/s and in the middle between 62 and 90 km/s. On the right are stars for which this is more than 90 km/s (the scale here is a factor of two smaller, so the 62 km/s circle would be twice as small as in the other panels). The important finding was, that stars that have space velocities less than 62 km/s move in all directions, but as soon as that value is exceeded the directions are suddenly limited to about half. Oort also investigated other samples of stars, but they only confirmed what he had already found, namely always such a sudden transition around 65 km/s.

Oort had no idea how the high-velocity stars fitted into the picture of the structure of the Sidereal System. It was clear to him that they could not be part of the Kapteyn System and had to be intruders with velocities higher than the escape velocity.

This work by Oort was the conclusion of his university studies. He had fulfilled the requirements for the doctoral (masters) examination, which he passed on November 5, 1921, again *cum laude*, two out of forty that year in his Faculty.

Fig. 4.0 Oort as a student wearing a so-called 'corps hat'. This was worn at special occasions by members of the fraternity, the Groningen Student Corps *Vindicat atque Polit*. From the Oort Archives

4

Via Yale to Leiden

*Make sure you work at a place which gives you opportunities to
learn, travel, explore, interact with intellectuals and develop new work skills.*
Abhishek Ratna

*I think of Princeton as being lazy and good-looking and aristocratic —
you know, like a spring day. Harvard seems sort of indoors —
··· And Yale is November, crisp and energetic.*
Francis Scott Key Fitzgerald (1896–1940)

A few months before Oort's doctoral exam (masters) on November 5, 1921,
Kapteyn had retired and Pieter van Rhijn had taken his place as professor of astronomy and director of the Astronomical Laboratory, which in the meantime
had been named after Kapteyn. Van Rhijn had appointed Oort as an assistant,
which had caused Oort problems because at least two more advanced students,
Jan Schilt and Egbert Kreiken, would have been passed over. Oort wrote to his
mother on April 30 of that year (1921):

> Yesterday morning, Professor van Rhijn asked me to his office and offered
> me to be appointed as his assistant. I then told him my problems with this.
> Objection Schilt, which has now been multiplied by a similar objection Kreiken
> (because he had failed the health test for the Indies) was of no relevance to him
> because my appointment probably would be long-term anyway. He would try,
> if at all possible, to make it a well-paid job (so no assistantship), so that I would

---

A. Ratna is an Indian marketing manager and management coach [44].

P. C. van der Kruit, *Master of Galactic Astronomy: A Biography of Jan Hendrik Oort*,
Springer Biographies, https://doi.org/10.1007/978-3-030-55548-1_4

Fig. 4.1 The Sterrewacht Leiden as it appeared as illustration on the title page of the *Annals of the Leiden Observatory* (parts 1 to 9; 1868–1902). From the archives of the Sterrewacht Leiden

stay longer and would not e.g. for financial considerations want to become a teacher. In principle, I have accepted this fine offer, which I consider to be a small honor, but I would first like to talk to you about it extensively, which he understood very well, and I am still completely free to either I accept it or not.

Of course, Oort's parents had no objection and the appointment was made. This meant that Oort would mainly do astronomical research (and assist with the teaching), which would eventually lead to a PhD thesis. But there were developments in Leiden, which resulted in a different course of events. In order to clarify that situation, we need to look into the history of the Sterrewacht Leiden (Fig. 4.1).

4.1   Leiden Observatory (Sterrewacht Leiden)

The observatory dates from 1633 when a quadrant was purchased and installed on the roof of the Academy Building for Willebrord Snel van Royen or Snellius (1580–1626). By means of triangulation of church towers, Snellius measured the Earth's circumference. With one important exception, Christiaan Huygens

(1629–1695), the discoverer of the rings of Saturn and its moon Titan in 1655, astronomy in Leiden remained rather insignificant. The appointment of Frederik Kaiser (1808–1872) as director in 1837 (he had been an observator since 1826) marked the turnaround. He managed to raise the money required for the construction of a real observatory with telescopes and offices, modeled on Pulkovo Observatory at St. Petersburg; the costs had been 100,000 guilders (currently equivalent to just below a million Euros purchasing power). The Observatory was inaugurated in 1861 (see [45–47] or Chap. 2 in [35] for an extensive historiography of the Sterrewacht, as Leiden Observatory became known, and [48] for an excellent treatise on Kaiser).[1]

Under Kaiser the Sterrewacht flourished as a center of high precision astrometry, which is concerned with determining positions of stars. Under his successor Hendricus van de Sande Bakhuyzen this tradition was continued, although the latter had rejected implementing new developments, such as photography in particular. In the 1890s Kapteyn hoped to get a photographic telescope for Groningen, but then van de Sande Bakhuyzen decided to refuse to give up the leading position of his observatory and managed to obtain it for Leiden. In 1908 van de Sande Bakhuyzen retired.

Initially the position had been offered to Kapteyn, obviously because he now was the most prominent astronomer in the Netherlands. But Kapteyn turned this offer down. The implementation of his *Plan of Selected Areas* had just begun and in that year he also made his first trip to Mount Wilson Observatory and Pasadena (at the foot of the mountain, where the observatory offices are) as a Research Associate of the Carnegie Institution. Not a very opportune time to take over the leadership of the Sterrewacht Leiden. Kapteyn however made a strong push for his assistant de Sitter (Fig. 4.2).

Willem de Sitter (1872–1934), born in Sneek, came from a lineage of lawyers, but broke with that tradition and studied mathematics in Groningen. He participated in Kapteyn's research program as a doctoral (for the masters) student by measuring up photographic plates. In 1896 he met David Gill, the director of the Royal Observatory at the Cape, who visited Groningen, and who was so impressed by de Sitter, that he invited him to come to Capetown to do his PhD thesis research. After consultation with his parents, de Sitter indeed left for South-Africa in 1897. In Capetown, de Sitter married Eleonora Suermondt (1870–1952) in 1898. The instrument that he would use for photometry (brightness measurement) of stars was not ready, so he did many accurate positional measurements of the four large satellites of the planet Jupiter. These were discovered in 1610 by Galileo Galilei (1564–1642) and played an important role in the acceptance of the heliocentric worldview,

---

[1] Unfortunately all these sources are in Dutch.

Fig. 4.2 Willem de Sitter. The instrument at the telescope is a Zöllner photometer, designed by Johann Karl Friedrich Zöllner (1834–1882) in 1858 and widely in use for many decades. The Bunsen burner on the left produces a stable flame that is projected into the field of view so that it can be compared with a star. In between the burner and the telescope are two polaroid plates that can be rotated with respect to each other to adjust the brightness of the flame so that the magnitude of the star can be determined. This picture comes from the *Album Amicorum*, presented to H.G. van de Sande Bakhuyzen in 1908 [36]

because the satellite system of Jupiter looked like a miniature planetary system. After his return to Groningen, de Sitter wrote a dissertation on the satellites of Jupiter, with which he obtained his PhD in 1901 as Kapteyn's first PhD student. De Sitter then concentrated on further studies of this satellite system – as he would do all his life – but also contributed to the processing of the plate material that Kapteyn regularly received from Donner in Helsingfors.

Like the planets in our planetary system, Jupiter's satellites cause small disturbances in each other's orbits. They move in orbits that are in 'resonance', i.e. the orbital periods relate with a high degree of precision to each other by small numbers . The first three, Io, Europa and Ganymede, have orbital periods in a ratio of 4 : 2 : 1. As a result, the disturbances are not random, but they recur systematically and regularly over time. Then those orbits are, as it were, trapped in these resonances, which gives a high degree of stability. De Sitter studied the additional effects of the fact that the orbits are not exactly circular and not in exactly the same plane, as well as the influence of the fourth satellite, Callisto, which has a resonance of 3 : 8 with Ganymede. Although de Sitter was strongly involved in observational studies, he had a theoretical inclination. He later worked with Albert Einstein (1879–1955) on models of the Universe based on the latter's general theory of relativity, resulting in particular in a model that became known as the 'Einstein-de Sitter Universe'.

A complication was that the brother of the director, Ernst Frederik van de Sande Bakhuyzen (1848–1918), also worked as an observator in Leiden at the Sterrewacht, and he also wanted to be considered for the directorate. In the end Ernst van de Sande Bakhuyzen was appointed director and Willem de Sitter professor of astronomy. Van de Sande Bakhuyzen was appointed extraordinary professor. However, he was a very conservative man (he had special permission to continue using oil lamps because he could not get used to electric light) and far from an inspiring leader. The telescopic research essentially came to a standstill. Shortly before his retirement in 1918, he died quite suddenly. With the support again of Kapteyn, de Sitter was chosen and appointed as the new director. De Sitter immediately carried out a fundamental reorganization of the Sterrewacht, in which Kapteyn actually had an important share.

The new structure of the Sterrewacht was that there would be three separate departments with deputy directors in charge of them and with delegated authority, overseen by the director who had the final say in everything. The Theoretical Department was to be led by de Sitter himself. Next, there was a 'Fundamental' Department, which would continue the program of astrometry and position measurements, and an Astrophysical Department, which focused on studies of the structure and evolution of stars. For the latter there was a good candidate in the person of Ejnar Hertzsprung (1873–1967) (Fig. 4.3). At that time, Hertzsprung worked at the Potsdam Astrophysikalisches Observatorium. He now is best known best for the Hertzsprung–Russell diagram, the fundamental diagram of the systematics of stars, which I will discuss later. He was Danish by origin, and ended up in astronomy via a detour. Karl Schwarzschild, director in Potsdam, had urged Kapteyn to introduce him to George Hale, so that he could work for some time at the Mount Wilson Observatory. Kapteyn had agreed and Hertzsprung came to Groningen for some time before Kapteyn would take him along on one of his visits to Mount Wilson. Hertzsprung and Kapteyn's second daughter Henriette Mariette Augustine Albertine (1881–1956) were engaged to marry before their departure. It would not be a happy marriage, according to many because of the not very

Fig. 4.3 Ejnar Hertzsprung as a young man (left) and Antonie Pannekoek. Hertzsprung's photo has been made available by the University of Aarhus [49]. Pannekoek's comes from the *Album Amicorum*, presented to H.G. van de Sande Bakhuyzen in 1908 [36]

social workaholic Hertzsprung, and in 1923 they chose to live separately (after a formal divorce in 1937 Henriette remarried).

The person that de Sitter and Kapteyn chose for the leadership of the Fundamental Department was Antonie Pannekoek (1873–1960) (Fig. 4.3). He had studied astronomy in Leiden, where he obtained his PhD in 1902. He became a convinced communist and Marxist, and moved to Germany. When the First World War broke out he happened to be in the Netherlands and could not go back, so he became a teacher. De Sitter selected him in 1918 for the position of deputy director. This looked to be case shut and done, because the liberal Prime Minister Pieter Willem Adriaan Cort van der Linden (1846–1935) had no objection, but then elections were held that resulted in a new, Catholic Prime Minister, Charles Joseph Marie Ruijs de Beerenbrouck (1873–1936), who ruled Pannekoek's socialist sympathies to be unacceptable. The Municipal Council of Amsterdam, however, was much less prejudiced and had no objections; so Pannekoek was appointed to the Amsterdam Municipal University, where he would eventually become a professor. For a more extensive discussion in Dutch of the reorganization of the Sterrewacht Leiden in 1918, see two articles by David Baneke [50, 51].

The astrometric position was briefly filled by Kapteyn as a part-time appointment after his retirement, but remained open after his death. Kapteyn had written recommendations to de Sitter for both Schilt and Oort; the latter in particular seemed to him to be an excellent candidate for an appointment in Leiden. De Sitter agreed but felt that Oort should learn more about astrometry first.

4.2   The IAU in Rome

The International Astronomical Union IAU is the worldwide organization that represents and coordinates all astronomical interests (see Box 3.1). Shortly after the First World War, international organizations of this kind were set up. It was the case, however, that the defeated nations of World War I (such as Germany and Austria) were excluded from such organizations. The irreconcilable tone of the manifestos of the time is extreme. Not only were these countries excluded from the League of Nations, which was set up to facilitate dialogue and prevent wars, scientific organizations also excluded them from membership.

Box 3.1 The International Astronomical Union

The International Astronomical Union (IAU) was founded in 1919. The mission is to promote and protect the interests of astronomy though international cooperation. The individual members are professional astronomers from all over the world. The IAU currently has more than 13,000 members from 107 countries. Of these countries, 82 are members as a nation.

The IAU's scientific and educational activities are divided into divisions, committees and working groups. The IAU secretariat is at the Institut d'Astrophysique in Paris.

The main activity of the IAU is the organization of scientific meetings. Each year the IAU sponsors a number of international symposia and a General Assembly every three years, at which six symposia take place in addition to meetings of divisions, committees, working groups, often jointly on a specific subject. The most recent one was in Vienna in 2018. The other activities are establishing values of fundamental astronomical and physical constants, unambiguous astronomical nomenclature, promoting educational activities and organizing informal discussions on, for example, the possibilities of large international facilities. The IAU is also the body that establishes names of astronomical objects or structures on their surfaces.

The IAU cooperates with organizations such as UNESCO to promote astronomical education and research in developing countries.
Adriaan Blaauw has written a book about the first fifty years of the IAU [53]. Recently, as part of the centennial celebrations, a new, comprehensive volume has been published [54]. The IAU website is www.iau.org.

In October 1918, the British Society of London organized a meeting with representatives only of the victorious countries (United Kingdom, United States, Italy, France, Belgium, Serbia and Brazil) on the creation of an umbrella organization of scientific academies. The proceedings were published in an astronomical magazine [52]. The tone is clear from the statement that 'unlike after previous wars, personal relationships between scientists from the Allied Countries and the central empires will be impossible for a long time to come'. The text leaves no doubt:

> War is inevitably full of cruelties [...]. These are not the acts we refer to, it is the organized horrors, encouraged and conceived from the beginning, with the sole aim of terrorizing inoffensive populations. The destruction of numberless homes, the violence and massacres on land and on sea, the torpedoing of hospital ships, the insults and tortures inflicted on prisoners of war, will leave in the history of the guilty nations a stain which the mere reparation of material damages will not be able to wash away. In order to restore confidence, without which all fruitful collaboration will be impossible, the central empires will have to repudiate the political methods the practice of which has engendered the atrocities which have roused the indignation of the civilized world.

In July 1919, the International Research Council (IRC) was established, and Germany was excluded from membership. Countries that had remained neutral during the war, such as the Netherlands, were allowed to join later, but it was ensured that the statutes were already complete and could no longer be changed if the neutrals still wanted to admit Germany, Austria and other former enemies. Kapteyn and his Groningen friend Gerardus Heymans (1857–1930), professor of philosophy and psychology, strongly disagreed with what they saw as a disastrous turn of events. In 1919 the Royal Netherlands Academy of Arts and Sciences (KNAW) wanted to join this IRC, and they fiercely resisted. They wrote an open letter *To the members of the academies of the allied nations and of the United States of America*, in which they described science as the 'the great conciliator and benefactor of humanity' [55]. Kapteyn's opinion also put him in a difficult position in relation to many colleagues at home and abroad, who were very much in favor of the exclusion of Germany, especially the influential British astronomer Herbert Hall Turner (1868–1938) of Oxford. But Kapteyn also had supporters, such as Svante Elis Stömgren (1870–1947), director of the Copenhagen Observatory. The latter worked to ensure that Germans had access to scientific material, such as in 1919 when others refused to share observations of a comet with German astronomers. 'Even my boys of 9 and 11 would never do such a thing', he said. Kapteyn was asked to join the board of the German Astronomische Gesellschaft; he agreed, not because he aspired to such functions, but because he might be able to help repair the rift. Kapteyn also found the British Arthur Stanley Eddington (1882–1944) at his side, who was the only Brit to attend the congress of the Astronomische Gesellschaft in 1920. Arthur Eddington carried a lot of weight; after all, during a solar eclipse in 1919, he had verified the bending of light of stars by the gravity of the Sun as predicted by Einstein.

The International Astronomical Union IAU was founded in 1919. Germany again was excluded from participation and membership. Kapteyn did not have anything positive to say about the IAU. The first General Assembly of the IAU was held in Rome in May 1922. The Dutch 'Committee for the IAU', which coordinated matters nationally in the Netherlands and was present in Rome, consisted of de Sitter, Hertzsprung, Albertus Antonie Nijland (1868–1936), professor of astronomy in Utrecht and director of the observatory there, and Jacob Evert Baron de Vos van Steenwijk (1889–1978) as secretary. The latter had a PhD in astronomy. He came from a family of politicians and after having been a teacher for some time he became mayor of Zwolle and later Haarlem and commissioner of the Queen in the province of North Holland. However, he remained actively interested in astronomy. Later he was also President-Curator of Leiden University. They were not the only Dutch participants, van Rhijn from Groningen was there too.

Oort was far too junior to be invited to this meeting. However, he was actually present, no doubt with the help of de Sitter and van Rhijn. Oort later remarked that he spent the Easter holidays (but that must have been the Whitsun/Pentecost holidays, given the dates) in Rome visiting a friend from university who lived there for a year. Indeed, Oort is on the group photo of the General Assembly (see Fig. 4.4). It was during this meeting that de Sitter spoke with the American astronomer Frank Schlesinger (1871–1943) about Oort. Schlesinger was director of the Yale Observatory, the observatory at the university of the same name in New Haven, Connecticut, and a leading astrometrist. The idea was that Oort would work at Yale for some time and learn the tricks of the trade there, before coming to Leiden. De Sitter and Schlesinger were good friends and the agreement was quickly reached.

Frank Schlesinger had started his career at Yerkes Observatory in Williams Bay, Wisconsin, of the University of Chicago, Illinois. This observatory was

Fig. 4.4 Detail from the group photo of the General Assembly of the IAU in Rome in 1922. Oort can be seen bottom right half hidden behind Henry Norris Russell. Straight above Oort Ejnar Hertzsprung and at the far left Frank Schlesinger. From the Oort Archives.

founded by George Hale in 1897 with funds from businessman Charles Tyson Yerkes (1837–1905) and Hale had built the largest telescope in the world with a 40 in. (about 1 meter) objective lens. While Schlesinger was at Yerkes, he had improved the techniques of parallax measurements with photographic methods. In 1905 he had moved to the Allegheny Observatory of the University of Pittsburgh, where he became director, but in 1920 he changed this position to that of director of Yale Observatory. The reason for this was the following. We already saw how between 1849 and 1863 Friedrich Argelander of the Bonner Sternwarte had produced the *Bonner Durchmusterung (BD)*. The sequel, the *Astronomische Gesellschaft Katalog (AGK)*, had been started in 1861 as an extension using meridian circles on fainter stars (up to magnitude 9, more than ten times fainter than the naked eye can see). This concerned about 200,000 stars. Nine observatories, including Leiden, were involved. The last part was published in 1921, but most of it was available earlier. Schlesinger now wanted to repeat this with wide-angle cameras to determine proper motions. He next planned to measure parallaxes and proper motions in the southern hemisphere in the same way as an extension of the *Cape Photographic Durchmusterung* of Gill and Kapteyn, and it was Yale University that was willing to make the required funds available for this. Schlesinger's reputation was great, as illustrated by the fact that he had been elected President of the American Astronomical Society from 1919 to 1922. The ideal person for de Sitter to send Oort to.

Oort had apparently not stayed until the end of the IAU meeting in Rome, because de Sitter wrote to him from Rome on the last day (May 10, 1922):

Amice,

I have to write to you right away about a very important development. I will
come straight to the point. Prof. Schlesinger of Yale Observatory offers you to
join his observatory for one year. The salary will be $ 900 + a room in the
observatory. You will have to make all your time available to the observatory for
10 months, while 2 months of the year are holidays. You shall start on the 1st
of September or the 1st of October at the latest. I propose that you come and
talk to me about it as soon as I get back to Leiden.

In a follow-up letter dated June 7, de Sitter clarified that the requirement
that Oort would have to work for the observatory 'a certain number of hours
per day + a certain number of hours per clear night' and that certainly some
time would be left for his own work. De Sitter compared the salary with
that of an observator at the Sterrewacht (4000 to 5000 guilders). Schlesinger's
$ 900 corresponded at that time to about 2,500 to 3,000 guilders, which was,
he felt, not unreasonable for a young astronomer who did not have a PhD yet.
Time was running out and Oort had to decide quickly. On June 8 he sent a
telegram from Groningen to de Sitter in Leiden with the text: 'Decided for
America. Preferably October. Oort.'

Also on June 8 Schlesinger wrote to de Sitter that he had submitted the
proposal for the appointment to the authorities of the university and that if
it was approved he would send a telegram with the text 'Confirmed'. And as
for the details, he confirmed that the salary would be as mentioned previously,
and:

> It is understood that the appointment is for one year, but if both parties are
> willing it may be renewed. Mr. Oort's duties will consist of assisting with the
> automatic zenith tube for determining the latitude. I will see to it that he gets
> as large a variety of experience as will be feasible at our Observatory in one year.
>
> The room at the Observatory to which he will be assigned is a large one and
> we will install a comfortable sleeping cot. If he wishes to occupy the room as an
> office and sleeping quarters, he will be expected to provide his own bed clothes
> including blankets, and will be expected to keep the room looking like an office
> in the daytime.
>
> I was well impressed with the little I saw of Mr. Oort at the meeting; and
> especially in view of your high recommendation of him. I am looking forward
> to becoming better acquainted with him and to have him work with us next
> year.

On June 12, de Sitter wrote to Oort that he had received the telegram with
'Confirmed' from Schlesinger. Through de Sitter Schlesinger wrote to Oort:

> I have already stated in my letters to Professor De Sitter how great a satisfaction
> it is to me to know that you are to be with us for a year. I am looking forward

with pleasure to seeing you early October and to your working with us during the year.

Oort's decision to go to Yale was profound and taken at a time when two matters of deep significance and sadness had come to him. On the same day that Oort sent his telegram to de Sitter, June 8th, his brother Hein died. He suffered from diabetes. At that time that was a deadly disease. The insulin treatment of human patients was first applied in January 1922 by Frederick Grant Banting (1891–1941), and although the use spread very rapidly—Banting was awarded the Nobel Prize for the discovery of insulin already in 1923—it came too late for Hein Oort. He had come to the conclusion that life was no longer worth living for him, and had traveled to Vienna. His parents went looking for him there and were there when he died. His mother sent a letter to the other children to inform them. His death left a deep wound in the Oort family.

And on June 18, 1922, Kapteyn died. This came of course as no surprise, but this too, must have come as a shock to Oort. Schlesinger wrote to de Sitter on July 26:

> Although you and Hertzsprung had prepared us for the worst with regard to Kapteyn's health, the news of his death came as a shock. The world has lost its greatest astronomer. Personally we feel we have lost one of our best friends.

Fig. 4.5 The Oort family, probably in 1922, shortly after Hein's death and before Oort's departure for the United States. Oort is on the right and his brother John in the middle. Next to John sister Emilie and his sister Jetske sits on the left. From the Oort Archives

Fig. 4.6 The SS Rotterdam (IV) of the 'Holland-Amerika Lijn', with which Oort traveled to the United States in 1922. From Wikimedia Commons [56]

In September, Oort left for Yale; the picture in Fig. 4.5 was probably taken shortly before that. The family doesn't look very happy.

4.3   To America

Oort left Rotterdam for New Haven and Yale Observatory on September 13, 1922 with the SS Rotterdam (Fig. 4.6) of the NV Nederlandsche-Amerikaansche Stoomvaart Maatschappij, better known as the 'Holland-Amerika Lijn' (HAL). The passenger list (kept in the Rotterdam Municipal Archives) shows that Oort traveled second class and had booked his ticket in Groningen for an amount of 350.50 guilders (current value approximately 2600 €). In the private part of the Oort Archives resides a travel log with a list of ten addresses. This refers to readers, who after reading should forward it to the next one on the list. At the top are several housemates from the Brugstraat in Groningen, then some family and the last the parents Oort. The travelogue starts on Friday September 15, 1922.

> It is so beautiful outside now. You cannot easily imagine yourself being on this big passenger steamer with bright lights all over it now sailing at full speed on the Atlantic Ocean. The sky is dark, sometimes with a small opening between the clouds and then you can see a few stars and then again it is dark. The sea

forms a big disk around us and we have left already quite some horizons behind us since we saw land for the last time yesterday afternoon, which consisted of a tall lighthouse standing on a small island off the coast of south-west England.

On Tuesday September 25, Oort wrote in it from New Haven.

The voyage over the ocean was in most respects extraordinarily pleasant, even though the weather was not really good. It started the afternoon of the first day when we entered Boulogne. The boat of the pilot that came to our side, sometimes disappeared completely under the waves. Dinner was served with care and I did not eat much. To bed early and I fell asleep soon and did not notice the storm that we had that night in the Channel. Next morning I found to my horror that the water almost ran over my bath, because the ship tilted so much and I did not feel very well until we reached Plymouth but it did not get worse. [...]

We had dancing parties two evenings on the deck, which were not so nice since few people danced and there were quite a few annoying persons. The presence of unpleasant people was one of the less nice points of the trip, especially later when I knew them by sight. You cannot escape people, in particular when the weather is poor and you cannot sit outside. I have sometimes looked with jealousy at people in third class. Among them there were quite a few interesting persons, at least at first sight. One of them was an old lady, who every morning when the weather was fine sat on a beam with a fat book with Russian characters, probably a Bible. And there were a few very pretty Slavic young ladies with colorful scarfs and also a fair number of expensively dressed people that seemed out of place and appeared to belong really to the upper class. They did all sorts of games there like hide and seek or tag and the poshest and the poorest participated with great enthusiasm. I think I will take third class on the way back, but maybe the people will be different then. First class is beautiful with many lounges, dining rooms, music chambers, reading rooms, etc. and many promenade decks. There is plenty of room for everyone. I have been there only twice. They were very strict in that respect and the get-togethers with the Hamburgers and Liesje Storm van Leeuwen had to take place at a staircase or across a fence. They were not even allowed in second class. [...]

On Friday morning we arrived in Quarantines at Sandy Hook still before New York and had a terribly long doctor's examination so that we only disembarked around noon and then still had to wait for luggage inspection. I had not slept very well that night and was very tired. On top of that I was suddenly all by myself in the large barrack of the H.A.L.; the Hamburgers had left already and Liesje S.v.L. with whom I was going to spend a day in N.Y. and then see her off, found out that her brother had met the ship and they had to leave that same afternoon. In spite of this I very much enjoyed my first moments in N.Y.!

The 'Hamburgers' refer to Hartog Jacob Hamburger (1859–1924), professor of physiology and histology in Groningen. He was also chairman of the (now Royal) Natural Sciences Society, in which Kapteyn was also very active. Their annual report shows that Hamburger was going to give a number of lectures at Johns Hopkins University in Baltimore and then made a lecture tour of the United States. Liesje Storm van Leeuwen (Storm comes from 'Stem') was Alide Marie Catherine Storm van Leeuwen (1889–??). It is unlikely that Oort knew her from school or college, because she was 11 years his senior. It appears that her eldest brother Willem Storm van Leeuwen (1882–1933) was a physician who was appointed professor of pharmacology in Leiden in 1920, where he specialized in allergic disorders, particularly asthma. Oort's father must have known him and Willem and he must have brought them into contact when it turned out that they were traveling on the same ship. The brother who picked her up was Arnold Storm van Leeuwen (1885–1948). He lived in Woodstock and Liesje was also on her way there. I will get back to this. The travel report continues as follows:

New York is like alcohol, certainly the first time you visit. Especially the pace and the noise add to the excitement. All traffic moves very fast and then stops suddenly when signs change, because Broadway and Fifth Avenue have sign posts such as our trains do. They do use red and green lights or big signs at almost every intersection. It is a lot of fun to watch all that action from a bus. [...]

In the afternoon [the second day] I left for New Haven, which is 130 km N.E. of N.Y., and is connected by electric trains. You go uninterrupted through suburbs [Oort called them villa villages] along the coast. New Haven itself looks like a beautiful town and in that respect it is an exception to most American cities, although I have no real experience with that. A large part of the center is occupied by university and student buildings, of which a very large number exist. There is for example a beautiful, large building, in old Gothic style, that has just been completed, with a large number of small stained glass windows and even artificial sagging roofs and broken windowsills! In spite of that I think it is a beautiful building; I will later send a photograph. It consists of various departments and low gates provide entrances to even smaller, cosy courtyards separate from the larger campus in the middle. I heard that some lady had donated 6 million dollars for it (to compare with the 1 million dollar that had been provided for the Peace Palace [in Den Haag]). I saw a room in it where the son of the Director of the Observatory lives, also decorated in old style, open fireplace, wide windowsills where you can sit, and wooden walls, which all are very beautiful. The luxury and excessiveness of it all, especially as far as the students are concerned, is really amazing. But then Yale is one of the most expensive universities. I will have my dinners in a large student-house and expect to learn more about all this and will some time write about it. But for the moment this: the majority of the students work for some of their income and there are only a few that get all their study money from their parents. There is a special office in the University, where you can go to look for a job.

Yale University is the third oldest institution for higher education in the United States; it was founded in 1701 (Harvard is the oldest and has been in existence since 1636), first as a college and since 1887 as a university. The name comes from Elihu Yale (1649–1721), born in Boston but raised in England. He was president of a large trading company (the British East India Company in Madras, India). After an appeal for financial support he sent a large donation to New Haven in 1817, including books and other goods. The latter were sold and the considerable sum of money it raised was used for a new building, which subsequently became Yale College. And that name was transferred later to Yale University. Yale is indeed an expensive university; it is part of the Ivy League, which is actually a competition between sports associations of eight colleges. It now includes some of the most prestigious American universities, including Harvard and Princeton. By the way, the name Yale has nothing to do with the famous brand of locks. The pin-tumbler lock was invented by Linus Yale Jr. (1821–1868), who established the factory of the same name in Stamford (coincidentally also in Connecticut), but that has nothing to do with the university. On the other hand, Linus and Elihu Yale do descend from the same Welsh family.

The travelogue of Oort ends on 27 September 1922:

> The Maverick, Woodstock (N.Y. State)
> Around midnight. You will never guess from where I am writing this. My bed is bordered by 4 still unpainted poles that have recently been sawed off. In the glow of a paraffin lamp I can only see with difficulty what I wrote and also what the room looks like: rough wooden walls, hardly any furniture, only a few beautiful carpets. Outside there is forest all around and the squirrels (a sort of small 'eekhoorn') make a lot of noise.

Oort had travelled to this place even before he was due to take up his post on October 1. It was the place where Liesje Storm van Leeuwen had gone and where originally Oort would accompany her, had her brother not picked her up. Now she had invited him to come and visit her, her brother and family. The trip was not without obstacles:

> My trip here was curious. I was not informed about the situation here and the way to reach it, so that, when I stepped off the train in Kingston at 7.15, it turned out that there was no connection to Woodstock, so I had either to walk 18 km or find a car. The latter turned out possible for $ 3.-, since there happened to be another passenger who also had to go that way. So, at high speed

we drove through the moonlit landscape (all major roads are asphalted here, so people can go much faster. In New Haven all streets, avenues and roads in the neighborhood are asphalted; I have seen no streets of gravel yet). Suddenly the driver turned into a narrow, rocky path, where he made the car jump over stones and moguls without any scruple. We went along like that for a while through a forest, until in the end we saw a wooden house, out of which a Jewess appeared who informed us where Miss Storm lived. She seemed delighted she had somebody to talk to, so the instructions were extensive. We drove on and at the place indicated there was a wooden house not too far from the path; when I entered I immediately heard Liesje's voice and to my surprise I found myself in the middle of 4 Dutchmen

Woodstock started as a village with Dutch people in the Catskill Mountains. In the early twentieth century, a colony was established in nearby West Hurley, where writers, musicians and artists lived in primitive dwellings. Since 1915 there were stage and concert facilities. At the time of Oort's visit there were mostly concerts of classical music during annual festivals. That is the same festival as the famous Woodstock festival of 1969.

The Maverick, where I am now, is on a lonely mountain slope, where they have built a number of small wooden houses, occupied in summer, mostly by artists that I am not too impressed by. The people with whom I stay, built these houses themselves. [...] The boy has a Ford, that he uses to get all kinds of things for us and takes us to all sorts of curiosities such as to a beautiful 'Fancy Ball' somewhere in the middle of the forest, where I danced with them and that was very nice.

And after the closing:

P.S. I returned on a boat along the Hudson river. Very pretty, although not as pretty as the Rhine.

4.4 Yale Observatory

Yale Observatory (called Winchester Observatory until 1920) was founded in 1880 (Fig. 4.7). William Louis Elkin (1855–1933), who was appointed there in 1884, did astrometry with a 6 in. heliometer. A heliometer is a telescope in which the objective lens at the top of the telescope tube is cut in two so that it is possible to move the two parts independently of each other. It is used to measure the distance in the sky between two stars by superposing their images and comparing the position of those two halves required to accomplish this. It

Fig. 4.7 Yale Observatory on an undated photograph, although it would have looked like this when Oort was working there. The dome on the right is that of the heliometer. Manuscripts & Archives, Yale University [57]

was first constructed for measuring the diameter of the Sun and that is where the name heliometer comes from.

In 1920 Frank Schlesinger (see Fig. 4.8) was appointed director. The reason why he came was that Yale University was willing to support his plan for an astrometric telescope in the southern hemisphere. But he was also interested in the problems of the variation of geographical latitude due to changes in the direction of the Earth's axis of rotation, not only the orientation in space, but specifically the position of the poles on the Earth's surface itself. And that is what Oort would work on. The positions of the poles change slightly over time, as a result of movements in the Earth's core and mantle (and nowadays also as a result of global warming causing the polar ice to melt; especially the ice on Greenland, which is not at the pole, contributes). One part is systematic and is called the Chandler wobble, which has a period of about 435 days and is related to the coupled circulations in the atmosphere and the oceans. The corresponding change in the position of the poles is no more than about 15 m on the Earth's surface. This may seem little, but it corresponds to almost three millionths of the Earth's circumference or a change in the orientation of the axis by about half a second of arc. And for astrometry, that is a lot. This 'latitude variation' was discovered in the 1880s by Karl Friedrich Küstner (1856–1936), when working on the Berliner Sternwarte (before moving to Babelsberg, Potsdam), but who later became director of the Bonner Sternwarte. Although the discovery is sometimes attributed to Seth Carlo Chandler (1846–1913)—after whom the above wobble is named –, he was in fact the first to confirm it.

Fig. 4.8 Frank Schlesinger at the 23rd meeting of the American Astronomical Society (AAS), in 1919 at the University of Michigan, Ann Arbor. He was president at the time. Source: University of Chicago [58]

The method, which Schlesinger and Oort used, goes back to Kapteyn. The director of the Geodetic Institute in Berlin, Friedrich Robert Helmert (1843-1917), had published in 1890 a short contribution on this effect in the German journal *Astronomische Nachrichten*. Results from various places gave different values for the change in the pole's position, which of course is not possible. Kapteyn had read this and wrote a letter back to Helmert, which Helmert, with Kapteyn's consent of course, had sent to the same journal for publication. In that letter Kapteyn proposed a more accurate method to measure the position of the pole. The method originally used was to measure the distance of a star from the zenith when is crosses the meridian; if, in a next measurement, the pole had changed position, then the geographical latitude of the observatory would be different, and then the distance of the star to the zenith would also have changed. Around 1850, George Biddell Airy (1801–1892), Astronomer Royal and director of the Royal Greenwich Observatory, used a bath of mercury as a mirror and then measured the position of the star on a photographic plate relative to the optical axis of the instrument, which by definition pointed to the zenith. However, Kapteyn suggested measuring *differences* in these zenith distances. If you expose a photographic plate for a long time, the stars on it describe circles around the pole. Now turn the plate 180° halfway through the night; in the second half of the night stars then describe circles curved the other way around. Stars from the two halves of the night are mirrored relative to the zenith. Choose two stars from different halves of the night. Now one coordinate of a star is the declination, that is the distance of a star from the equator. If you now know the difference in declination of those two stars, you can calculate declination of the zenith from the difference between the orbits of those stars on the photographic plate and from that the position of the pole and the geographical latitude of the observer.

Schlesinger used a 10 in. (25 cm) telescope, which was fixed on the zenith. Oort's task was to insert a photographic plate at the beginning of the night, as soon as the Sun had set start the exposure, turn the plate holder 180° in the middle of the night, end the exposure before sunrise and develop the plate. In principle, the plate can then be measured during the day, but in practice this process was lagging far behind.

It soon became apparent that the lens of the telescope unnecessarily limited the field of view and it was sent back to the manufacturer to improve it. But this caused a delay, since it was not back until August 1923. It was soon agreed that Oort would stay in New Haven for a second year. The staff was so small that the measuring of the plates was much delayed. After some time the work was done together with another student, Carl Leo Stearns (1892–1972), who actually continued it after Oort had left. Later, all the plate material was sent to Leiden, where it was measured by the 'calculator'[2] Gerrit Pels (1893–1966) with an instrument from the Amsterdam institute. But by then it was 1930. The outcome was disappointing; it turned out that the plate reversal in the middle of the night had not been done accurately because either the plate in the plateholder could shift slightly, or because some parts of that holder were not sufficiently rigid. This construction fault made all the material unusable, which was accepted definitely by Schlesinger only in 1935.

---

[2]Calculators, or 'rekenaars' in Dutch, were persons that performed arithmetic and sometimes mathematical tasks for the staff. Oort often had two or three such calculators working for him.

4.5   Life in New Haven

The Yale period was certainly not wasted time. Oort learned a lot about astrometry, but also gained much other experience. He did more work on his high-velocity stars and on some other projects, and visited astronomical congresses. Figure 4.9 shows Oort with two other young Dutchmen, Willem Luyten and Peter van de Kamp (1901–1995), who were both to remain permanently in the United States. This congress took place in December 1923 at Vassar College, Poughkeepsie, New York. It had been founded in 1861 by brewer Matthew Vassar (1792–1868) as the first institution of higher education for women that could award academic degrees.

But Oort also had an active social life. In particular, he had much contact with a local professor, Charles Cutler Torrey (1863–1956) and his wife Marian Edwards Richards (1877–1947). Torrey studied the relationship between Christian and Islamic religions (he also wrote a book on the Jewish basis of Islam) and translated the Gospels into Aramaic. He also worked as an archaeologist and was co-founder of the American School of Oriental Research, later the Albright Institute, in Jerusalem, of which he had been the first director (1900–1901). His wife was an author, best known for her book *My outrageous cousin*. They had a little daughter Anne (Nancy) Torrey (1920–2009).

In the *AIP-Interview* [6] Oort said that a cousin of his, who taught Eastern languages in Leiden, brought him into contact with the Torreys. He probably referred to Karel Hendrik Roessingh (1886–1925), professor of theology in

Fig. 4.9 Three young Dutch astronomers at the meeting of the American Astronomical Society in 1923 at Vassar College, Poughkeepsie, New York. On the left Willem Luyten, and on the bottom Peter van de Kamp. Oort is on the right in the picture [59]

Leiden since 1916, who was a cousin of an aunt of Oort. Even after his departure Oort remained in regular contact with the Torreys, whom he addressed as 'Uncle Charles' and 'Aunt Marian'. Through them, but undoubtedly also through the Schlesingers, Oort came into contact with many others. His diaries mentioned that he was regularly asked to have dinner with families and also had contacts with daughters of his age from those families (see Fig. 4.10). There seems not to have been any serious relationship, although there is one example, about which he wrote to his parents (on Sunday, February 24, 1924).

> As I wrote to you, I've become good friends with Doris Thompson. You would be a little shocked by her American manners and many of her American traits, but basically she is a sweet and sociable girl, and miraculously little spoiled by her good looks and popularity. She also differs from many other girls in that she seems to have a clever mind. She is certainly capable of distracting my thoughts completely away from my work and it is a great pleasure to visit her or go out with her. Monday evening we went skating in the moonshine on the pond in Woodbridge; the day before yesterday there was a big ball at the Lawn Club, for which she was on a committee and I saw her in the environment where she feels most happy and at home perhaps, every half minute a different dancer, so that it was not possible to dance more than ten paces with her.

Apart from this, she does not appear again in letters or diaries.

Fig. 4.10 Oort during his stay in New Haven. It is unknown who the two ladies in the picture are. They may be two of the three daughters of the Welch family, whose mother, according to Oort's diaries, provided him with biscuits and marmalade for his lunch at the observatory. From the Oort Archives

4.6   To Leiden

It was clear soon that Schlesinger was extremely satisfied with Oort. Already in a letter to his parents dated December 15, 1922, Oort wrote that Schlesinger had offered him a better-paid job, which would then be related to a new telescope. Oort would be involved in the Yale telescope, for which Schlesinger had plans to build it in the southern hemisphere. Oort was supposed to go to South-Africa for some time then. Indeed, in 1925 Schlesinger installed a 26 in. telescope near Johannesburg. This telescope was eventually moved to Mount Stromlo near Canberra, Australia, where it was destroyed in the bush fires of January 2003.

Leiden was not planning to let Oort go either and both Hertzsprung and de Sitter wrote him extensively about a future at the Sterrewacht Leiden. A period in Johannesburg would not hurt, according to Sitter, as long as it would be for a year or so after which he would then come to Leiden. The letter from de Sitter (March 7, 1923) was in English.

> Dear Jan,
> The other day your parents came here to speak to me about your plans and the question of your going to the southern hemisphere for Professor Schlesinger. I promised them to write you myself. I am doing this in English, so that you can, if you wish, show my letter to Prof. Schlesinger and talk it over with him.
> It may be that you are now to make a decision that will to a great extent be decisive for your future scientific career. I do not wish to influence you in making your decision, I only wish to aid you in putting the question to yourself as sharply as possible.
> I think, it comes to this in the end: do you wish to become a theoretical astronomer or an observer? In the first case you should try ultimately to come to Leiden, in the second you should stay in America, or in any other good climate. In *both*[3] cases a trip to South-Africa will do you a lot of good.
> An observing astronomer must be in a good climate, a theoretical astronomer must be in a scientific atmosphere. I know of no instances of great theoretical astronomers who have worked successfully away from general scientific surroundings. Of course Leiden has not a monopoly of the latter, but the position as Adjunct-Director at this Sterrewacht offers exceptional opportunities to a theoretical astronomer. By a theoretical astronomer I do not mean a mathematician: Kapteyn and Hertzsprung I consider theoretical astronomers.

In modern language, what de Sitter means would not be a theorist, but an interpretative astronomer. A theorist is the 'mathematician' de Sitter mentions. Indeed, Oort is not a theorist by today's standards, but an interpreter of

---

[3] De Sitter's typewriter must have had an option to produce text in italics.

observations. After all, Kapteyn had taught him to stay close to the observations at all times. De Sitter continued:

> For a theoretical astronomer in the wider sense I think the position as Adjunct-Director would be nearly ideal. As you know the Sterrewacht comprises of three departments: the fundamental, the astrophysical and the 'theoretical' (i.e. astromechanical) departments. The Fundamental Department includes, of course, the meridian-circle, but it by no means consists of this instrument exclusively. The fact that the Pole and the vertical are the only directions that can be fixed fundamentally lends to the plane through these directions a paramount importance, but there is a great many fundamental problems that can, and must, be solved by other methods. The Sterrewacht Leiden must not be fettered by tradition, if it thinks that these other methods are preferable over the old ones.
>
> The chiefs of the three departments, the two Adjunct-Directors and the Conservator (Dr. Woltjer) have a large degree of freedom. They have under them a staff of scientific assistants to whom the details of the execution of their plans can be entrusted. They themselves must be *leaders*, capable of choosing for themselves, though of course in consultation with the Director, the problems for their department and the means of solving those problems.
>
> As regards the Fundamental Department especially, I hope to be able to find an Adjunct-Director to whom I can leave the direction of this department almost entirely. I do not wish to exercise control over his work except in very broad lines: it must be really important fundamental work, not casual investigations, that can be done anywhere and by anybody.
>
> The post is at present unoccupied, as you know, by the death of Kapteyn. I cannot appoint you to it at once for two reasons. Firstly you are still a little too young to put you at the head of men who are much older than you. Secondly I do not know how you will develop, whether you will be a mathematician, or an astrophysicist, or a general theoretical astronomer. If you were a duplicate of either Hertzsprung or Woltjer I might have to try and find another man who more fulfills the demands of the case. I want a man who, to continue in the same stile [style], is a duplicate of Kapteyn or of Newcomb. I am willing to wait some time until I can get such a man, but I do not like to wait longer than is absolutely necessary, for at present all the work of the Fundamental Department falls on me, and makes a too heavy demand on my time.
>
> *If* your ambition lies in the direction of what I have called general theoretical astronomy, I honestly think you could not find anywhere a better position than the one I could offer you here, as soon as you are qualified for it. The best you can do in that case is to get as much and as varied experience as you can. You cannot have a better teacher as Schlesinger. As to your going south in charge of the big telescope, this would certainly be very valuable for your astronomical education, but I should say from this point of view one year of it would be sufficient, and you should also try to gain experience of other instruments and methods, including those of fundamental astronomy. Of course I do not know

whether Professor Schlesinger would send you out for one year only. I cannot promise to wait much longer than the beginning of 1925, i.e. until our present meridian program will be completed.

I say *if* your ambition lies that way. Whether it does no one but yourself can decide, and it would be foolish of me to wish to influence you. No good can come of constraint in such matters. Only I may be allowed to say that I hope it does, for in my opinion many important problems of the near future lie in the direction of fundamental astronomy.

I have in this letter spoken out my mind rather freely, and I hope you will consider it as confidential, and show it to nobody else but Professor Schlesinger.

Yours most sincerely,
[signed W. de Sitter]

On 25 March Oort wrote back that he had decided not to go to the southern hemisphere for 5 years and to be available for the position of deputy director. Schlesinger had agreed to this immediately after having seen de Sitter's letter. Indeed, Oort did not consider himself a mere observer. He also announced to come to the Netherlands in July for his military service. Oort had originally been called for initial training (boot camp) for military service in January 1920, but this had been deferred to complete his studies, a year at a time, eventually until the summer of 1924. Oort had told Schlesinger that he would have to serve in the initial training for a period of 2 to 6 months, depending on whether he would have to do part of his training as an officer. In March Schlesinger had written to the Dutch Minister of War that he wanted to offer Oort a position to go to South-Africa for several years to take the lead in building a large telescope and requested that Oort be limited for no more than two months. This was accepted.

According to his military passport, Oort entered military service on 23 July, 1923 and went on long-term leave on 30 August (Fig. 4.11). So he got a very good deal. I have not been able to find any record of his travel by boat to the Netherlands in the archives of the HAL in Rotterdam, so he must have used a different company. However, there was a record of his return journey as a second class passenger on the SS Veendam, which left Rotterdam for New York on October 10, 1923. In November 1923, Oort wrote to de Sitter that he intended to stay in New Haven until April (1924) and then travel across the United States for several months, before coming to Leiden in the summer. On February 1, he wrote that he had received the telegram offering the position and had returned a telegram with acceptance. On February 7, he confirmed that he agreed with the salary:

Fig. 4.11 Oort (front) during his military service in the summer of 1923. From the Oort Archives

> As long as I'm not married – and there is no plan to do so at this at the moment – the matter of the salary is not important to me.

The salary was 3600 guilders a year; that was about twice the salary of an ordinary worker, not very high, but Oort had not even obtained his PhD yet.

Oort bade farewll to the staff at Yale Observatory on May 7, 1924; in the days before and after this, according to his pocket diaries, he visited various addresses to say goodbye to the people he knew. He first travelled by train to New York and from there to Albany, New York, where he spent a few days with the director of the Dudley Observatory, Benjamin Boss (1880–1970), the son of the maker of the catalogue that Schilt and he had used for their prize-winning essay. Via Buffalo he continued his journey to Chicago and Kansas City, and finally to Flagstaff, Arizona where he arrived on May 23rd. He must have visited the famous Lowell Observatory, where Vesto Melvin Slipher (1875–1969) was the first to observe the rotation of an external spiral galaxy in 1914 (I will come back to this later). And he also descended on a donkey into the nearby Grand Canyon. He then went on, ending up in Pasadena, California, near Los Angeles, after admiring the beautiful deserts and yucca trees along the way. Already the day after his arrival (May 28) he was taken to the Mount Wilson Observatory, which was then the Mecca of optical astronomy. It had been founded first as a solar observatory by George Ellery Hale (1868–1938), but he soon exploited the potential for a nighttime observatory. After the 60 in. Telescope was put into operation in 1908—then the largest telescope in the world –, an even larger one was built very soon, with money provided by steel manufacturer John Daggett Hooker (1838–1911). This 100 in. Hooker telescope, again the largest in the world, was inaugurated in 1917 and was used by Edwin Powell Hubble (1889–1953) in particular to make spectacular photographic images of galaxies. From Oort's pocket diaries:

> ... in the observatory bus up (2000 m), where I meet M. [van Maanen]. Beautiful trip with wide views and steep abysses. The whole is very impressive and one is immediately taken by the whole, especially the location of the Monastery and the view from the balcony. Very good meals, all free, and everyone together, workshop men, bus drivers, janitors, night assistants, etc. See work on the 100″ mirror to remove astigmatism, then go to Monastery to finally write to Miss W. [one of the Welch daughters in New Haven?], and later again to 100″ and midnight lunch. Wonderful clouds beneath us.

Adriaan van Maanen (1884–1946) was a Dutch astronomer, who after obtaining his PhD in Utrecht (with Albertus Nijland, but with major supervision by Kapteyn with whom he worked in Groningen for some time) moved to Pasadena to work at Mount Wilson, where he stayed for the rest of his life. It was Oorts intention to work at the Mount Wilson Observatory and its offices in Pasadena for some time and he stayed there for almost two months. He talked extensively with van Maanen and other staff members who we will meet again later, including George Hale, Walter Sydney Adams (1876–1956), Gustav Benjamin Strömberg (1882–1962) and Edwin Hubble. He visited a number of times the brother of Charles Torrey, Joseph (1862–1952), who lived in Pasadena, and his daughter Helen Maria (1901–1990), with whom he went to the Egyptian theater (the first movie palace on Hollywood Blvd, which had opened in 1922). And with staff member Harold Delos Babcock (1882–1968) he went on a sailing trip to Balbao Island off the coast of Los Angeles. Oort always kept a special love for Pasadena and he visited it and the observatories (Mount Wilson but later also Palomar Observatory) as often as possible.

Oort left in July and first traveled on to the San Francisco area, also known as the Bay Area. There he visited Lick Observatory, where he arrived on July 28. Lick Observatory is located on Mount Hamilton near San Jose, about 80 km south of San Franciso. This observatory dates from around 1880 and was built with funds from James Lick (1796–1876), an investor, piano builder and landowner. Two well-known astronomers who worked there and who we will also encounter again later are Charles Donald Shane (1895–1983) and the Swiss-born Robert Julius Trumpler (originally Trümpler, 1886–1956). Figure 4.12 shows Oort at the Lick Observatory with those two astronomers and some members of the local staff. Oort then traveled by train via San Francisco, Berkeley, Denver, Detroit and Chicago to New York, where he arrived on August 10. After a day in New York he traveled on to Boston, where

Fig. 4.12 Oort visiting Lick Observatory in 1924. The two pipe-smoking gentlemen are Julius Trumpler on the left and on the right Donald Shane. The ladies are not identified. From the Oort Archives

he visited the Harvard Observatory and the university in nearby Cambridge (Massachusetts). To his regret, the director, the famous Harlow Shapley, was not present, but he had a lot of time to have extensive conversations with Willem Luyten.

Before he left for Europe, Oort took a few days holiday to visit the Torreys. They owned a second home on the Cranberry Islands, a group of islands off the coast of Maine, about 50 km south and a little bit to the east of the major city of Bangor. Apparently, he took a boat in Boston to get there. On the largest island, Great Cranberry Island, the Torreys had a house, where they often went on holidays. The picture in Fig. 5.0 was taken there.

For the return journey Oort also chose an unusual route. He first traveled to Quebec in Canada. His pocket diaries show that he had already planned and booked that trip in Pasadena. On August 22, 1924 he left with 'Mount Royal' (that must be the name of the ship, which is then named after the hill Mont Royal in the city of Montreal, which is also named after it). He traveled via England, probably arriving in Liverpool, at least he mentions that he traveled to London by train. The trip on the boat was quiet; he wrote that he read, sometimes played games and danced (sometimes it was fun, but there were also girls who could not dance so well, a Scottish young lady he describes as dancing 'like a frog'). In London he had very little money left, so he could not afford breakfast and on the boat across the North Sea he could not pay for a cabin. He finally arrived in Leiden on August 31, 1924. Just in time to take up his appointment at the Sterrewacht Leiden and university.

Fig. 5.0 Oort visiting the Torrey's on the Cranberry Isles in the summer of 1924, where they owned a holiday home. This was days before Oort started on the journey from his two years at Yale Observatory back to the Netherlands to take up his appointment in Leiden. From the Oort Archives.

5

Rotation and Dynamics of the Milky Way Galaxy

Look across the Universe, and you'll see that almost everything is rotating.
The Earth rotates on its axis as it orbits the Sun.
And the Sun itself is rotating. As you can probably guess,
we even have galaxy rotation with our Milky Way Galaxy.
Universe Today [60]

It is the harmony of the diverse parts, their symmetry, their happy balance;
in a word it is all that introduces order, all that gives unity,
that permits us to see clearly and to comprehend at once
both the ensemble and the details.
Jules Henri Poincaré (1854–1912)

When Oort returned to Leiden, his parents were in the Dutch East Indies for a year. Already on September 1, the day after his arrival, Oort was introduced to the staff at the Sterrewacht and started his appointment. His old study friend from Groningen, Jan Schilt, also worked in Leiden. Schilt successfully defended his PhD thesis in Groningen in 1924. He had developed a photometer with which one could measure the apparent brightness or magnitude of a star on a photographic plate by letting light shine through it and measuring the intensity of the light that came out on the other side, and compare this with

© The Editor(s) (if applicable) and The Author(s), under exclusive license
to Springer Nature Switzerland AG 2021
P. C. van der Kruit, *Master of Galactic Astronomy: A Biography of Jan Hendrik Oort*,
Springer Biographies, https://doi.org/10.1007/978-3-030-55548-1_5

the intensity if there was no star on the plate. For this he used a galvanometer that Frits Zernike had developed. Not much later, in February 1925, Schilt married Johanna (Jo) Timmer (1898–>1982) and then left for Mount Wilson Observatory and a year later for Yale University. Oort spoke extensively with Schilt and in the evening of his first day he was invited for dinner at de Sitter's home. On 3 September there was a meeting in Leiden of the Netherlands Astronomers Club NAC, the professional association of astronomers. Oort picked up van Rhijn at the railway station. On September 12, Oort took the train to Groningen to spend a few days with his old student friends and professors (Zernike and van Rhijn). But when he was back in Leiden, his work started immediately in the fundamental department (Fig. 5.1).

Fig. 5.1 Sterrewacht Leiden photographed from the south from a military aircraft in 1924. The canal is the Witte Singel. The two main telescopes (10 in. and 6 in.) are the two largest domes on top of the main building. The detached building on the right, the photographic department, has a dome which houses the 13 in. photographic telescope. In the main building from the left (west) we have the house of director de Sitter and the room for the meridian circle. On the right side of the main building two houses (at that time occupied by Hins and Hertzsprung). At the top of the photo the Academy Building and in between the Botanical Gardens. The freestanding cylindrical dome on the left contained several smaller instruments. From Wikimedia Commons [61]

5.1 Absolute Declinations

De Sitter (see Fig. 5.2) put Oort on the problem of absolute declinations. The position of a star in the sky is described, as we have seen (see also Appendix A.2), with two coordinates, of which the declination is the angle with the equator. I already mentioned that the measurement of declination was basically done with a meridian circle, where the declination is measured using the angle with the horizon that the telescope makes during meridian passage. The position of the pole can be measured with a circumpolar star, which is so close to the pole that it can also be observed when it passes through the meridian below the pole. The pole is then halfway between these two passages.

This sounds straightforward, but there are two effects that make this much more difficult in practice. In the first place, there is inevitably bending of the tube of the telescope under its own weight, so the star always seems to be a little higher above the horizon than it actually is. And secondly, there is refraction in the atmosphere. Light is deflected a bit by the air, and more so when the density increases, so the light travels on a curved path, so that again the star seems to be a bit higher. These effects are of course absent in the zenith, but increase as one gets closer to the horizon. To give an idea of the magnitude of the effect: at 45° above the horizon the refraction is of the order of a minute of arc, so far from negligible. The two effects work in the same direction, but although

Fig. 5.2 Willem de Sitter in the 1920s. From the archives of Adriaan Blaauw

astronomers knew of their existence, it was not possible to calibrate these effects, so one could not correct for them. There were considerable differences between declinations for a star in different catalogues (the values in the other coordinates, right ascension, measured by timing of meridian passage, agreed much better). As early as the 1880s Kapteyn had devised an ingenious method to circumvent this and measure 'absolute declinations'. The method is too complicated to describe here in detail. But to give you an idea: he used stars that went through the meridian almost at the same time at about the same altitude above the horizon, but on different sides of the zenith, one in the south and the other in the north. For his application he then only needed the difference between those two roughly equal heights and then the bending and refraction canceled. He also used times of passage through the primary vertical, which is just like the meridian but now from east through the zenith to west. His article on this ingenious method actually brought him in contact with David Gill. Although Gill started an observing program to do this from the Cape, it did not lead to actual results. Oort and de Sitter, both students of Kapteyn, were well aware of this work, but its execution was rather complicated.

The Dutch amateur astronomer, Carl Heinrich Ludwig Sanders (1864–1937), had after various other jobs owned a coffee plantation in Matube in Portuguese Congo (it is now in Angola), not far from the equator. He did observations in collaboration with the Sterrewacht Leiden, through director Ernst van de Sande Bakhuyzen, who also supported him financially after his plantation had gone bankrupt. It was known that it was possible to measure declinations free of those two effects from the equator by determining exactly where on the horizon stars rise and set. Consider Fig. 5.3, which illustrates the situation at the equator. North Pole N is exactly on the horizon. The path of

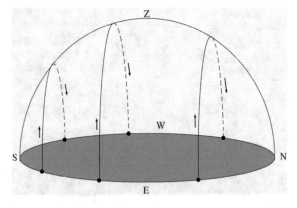

Fig. 5.3 The daily path of a star at a location on the equator. For further explanation see the text. Figure by the author

a star starts at the eastern half of the horizon, stars rise perpendicular to it and set on the western side (thick points in the figure). The angle between these two points on the horizon and the North Pole or half the angle between the two points (related to the azimuths) is exactly the polar distance (which is 90° minus the declination). One can measure that angle without being affected by bending and refraction, because it is measured along the horizon. Now Sanders had looked into this and derived how one could do this in case one was not exactly on the equator and in case one could not observe the star on the horizon, but a few degrees above it. Then corrections needed to be applied for this, but these are small and easily applied.

Soon Oort produced two papers. One was with de Sitter and in it they proposed to set up three special observation stations, one at the equator and two at about ±35° latitude. The one on the equator would then be equipped with a horizontal telescope (broken, i.e., with a prism so that you could look into it from above) for measurements of azimuth (angles along or parallel to the horizon). The other two would then use Kapteyn's method in a simplified form. The latter were never realized, but the equatorial method gave rise to two Kenya expeditions of the Sterrewacht Leiden, which I will discuss later.

The second article made use of measurements by Sanders to test his equatorial method. It indeed proved to work. Sanders returned to the Netherlands in August 1926 and was appointed senior assistant at the Sterrewacht. In the next chapter I reproduce a group photograph of the Leiden staff in 1931 (see Fig. 6.3), where Sanders can be seen as the somewhat heavy man, second from the left in the second row.

5.2   PhD in Groningen

Of course, it was very important for Oort to obtain his PhD degree. There was then the question who would be his 'promotor' (supervisor), Ejnar Hertzsprung in Leiden or Pieter van Rhijn in Groningen. In the end, the second option was chosen. After all Oort's research into the stars of high velocity had been started there, and this also was also a natural and suitable subject for a dissertation. Before I discuss that, I need to describe some background concerning the structure of our Galaxy.

It is instructive to use two figures, which Oort made himself, although later (see Fig. 5.4). They appear in a book by Willem de Sitter, *Kosmos*, based on a

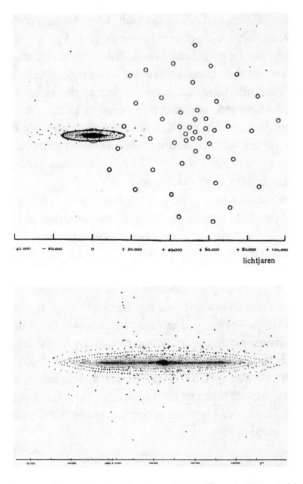

lichtjaren

Fig. 5.4 Illustrations attributed to 'Dr. Oort' used by Willem de Sitter in his book *Kosmos* [62]. At the top the earlier situation with Kapteyn's Universe located well away from center of Shapley's system of globular clusters. In both figures the horizontal scale is centered on the Sun and is at the top in lightyears and at the bottom in pc, at the top approximately from −13 to +33 kpc (kpc), at the bottom from −15 to +40 kpc. From the Oort Archives

series of lectures at the Lowell Institute in Boston in November 1931.[1] At the top the situation at the time of Oort's first investigations, with the 'Kapteyn Universe' removed far away from the center of Shapley's system of globular clusters. The Kapteyn Universe was much smaller than Shapley's system of globular clusters, while the center of the latter was in the Milky Way. The

---

[1]The Dutch version, published by Stockum, Den Haag (1934) re-published by Amsterdam University Press (see www.nl.aup.nl/books/9789089641540-kosmos.html), is available electronically via the 'Digitale Bibliotheek voor de Nederlandse letteren', see www.dbnl.org/tekst/sitt003kosm01_01/.

problem was the neglect of absorption (really mostly scattering) of light by interstellar dust. Shapley had found in globular clusters outside the Milky Way plane that there was no indication of any substantial effects in the colors of stars in the clusters. Because of this absence, absorption had been neglected; later it turned out that the dust certainly was there, but that it was concentrated in a thin layer in the plane of the Milky Way. As a result, the Kapteyn system was three times too small and the Shapley globular system twice too large (it is concentrated towards the center and there are relatively many clusters close to the Milky Way on the sky). At the bottom in Fig. 5.4 an impression of the new picture that had developed.

In reality, the Kapteyn System is only a part of a flat disk, which rotates. The globular clusters form a more spherical structure called the halo. The Sun and the stars near to us go around the center of the Galaxy at a velocity of over 200 km/s, but have relative, random velocities of only a few tens of km/s. In the halo the random velocities are often more than a hundred km/s, but there is much less systematic rotation (at most a few tens of km/s). So the system of globular clusters moves systematically with respect to us at about 200 km/s. Oort's high-velocity stars are part of the halo and the fact that they come from one hemisphere is nothing other than a reflection of the rotation of the disk. The reason that the directions of the velocities of Oort's stars do still cover about half the sky is because they have such great velocities among themselves.

Because of the dust in the disk, we cannot see globular clusters in the plane of the Milky Way. During work he did at Yale, Oort found that the high-velocity stars showed the same feature. So he concluded that there had to be a connection between those stars and the globular clusters. That gave him the idea that they were indeed 'aliens', as he called them, that had to come from outside the Kapteyn System.

Oort's PhD thesis was entitled *The stars of high velocity* (see Fig. 5.5) and was published in the series *Publications of the Astronomical Laboratory at Groningen*. The defense took place on Saturday, May 1, 1926. The most important conclusion was that these stars clearly come from one half of the sky and must therefore come from outside the Kapteyn Universe and that there is a clear and abrupt separation with stars whose spatial velocities are less than 63 km/s. Below this limit there was absolutely no deviation from symmetry in the directions of the velocities. This was an important point, because the Swedish astronomer Gustav Strömberg, who worked at Mount Wilson Observatory and whom Oort had met when he was in Pasadena, had found that such an effect, albeit much milder, also existed at lower velocities. Later it turned out that Strömberg was right, but that the abrupt and strong transition Oort had found remained. We now know that that transition at 63 km/s is the escape

THE STARS OF HIGH VELOCITY

PROEFSCHRIFT TER VERKRIJGING VAN DEN GRAAD
VAN DOCTOR IN DE WIS- EN NATUURKUNDE, AAN
DE RIJKS-UNIVERSITEIT TE GRONINGEN OP GEZAG
VAN DEN RECTOR-MAGNIFICUS Dr. J. A. BARRAU,
HOOGLEERAAR IN DE FACULTEIT DER WIS- EN
NATUURKUNDE, TEGEN DE BEDENKINGEN VAN DE
FACULTEIT IN HET OPENBAAR TE VERDEDIGEN OP
ZATERDAG 1 MEI 1926, DES NAMIDDAGS TE 4 UUR,

DOOR

JAN HENDRIK OORT,

GEBOREN TE FRANEKER.

GEBROEDERS HOITSEMA. — 1926. — GRONINGEN.

Fig. 5.5 The title page of Oort's PhD thesis. The text under the title reads: Dissertation to obtain the degree of Doctor in Mathematics and Natural Sciences at the University of Groningen on the authority of the Rector Magnificus Dr. J.A. Barrau, professor of Mathematics and Natural Sciences, to defend publicly against the objections of the Faculty on Saturday 1 May 1926, at 4 o'clock in the afternoon by Jan Hendrik Oort, born at Franeker. From the Oort Archives

velocity from the disk of our Galaxy at the position near the Sun. Oort decided that the high-velocity stars had probably escaped from the globular clusters, but the rotation of the disk had not (yet) occurred to him.

Figure 5.6 is from Oorts thesis and shows the distribution of the spatial velocities of the high-velocity stars. So these points are *not* their positions in the sky; Oort's stars were spread all over the sky. But it shows their spatial velocities. Shift these vectors, as it were, to the Sun and determine from which direction they point away. The horizontal scale is Galactic longitude with 0° on the left and 360° at the right. In the older notation used here 0° is one of the two nodes where the plane of the Milky Way crosses the celestial equator; nowadays we take the direction to the center of our Galaxy as the zero-point of Galactic longitude. In Fig. 5.6 the center is at longitude 327°, a bit to the left of

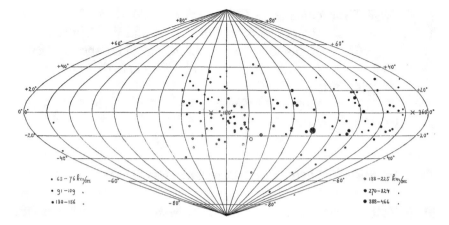

Fig. 5.6 First figure from Oorts thesis. It shows the distribution on the sky of the directions to which the high-velocity stars are moving. The horizontal axis is the plane of the Milky Way. From the Oort Archives

Fig. 5.7 Oort with his family and friends during the dinner after the defense of his dissertation on May 1, 1926. Next to Oort his fiancée Johanna Maria (Mieke) Graadt van Roggen. To the right at the table Willem de Sitter and his wife and standing behind them Pieter van Rhijn. From the Oort Archives

the right edge. The rotation of the Galaxy, as we now know, is 90° from there, which corresponds to longitude 57° in the figure, and that is where Oort's high-velocity stars then should be coming from. The figure shows where the velocities point *toward* and the center of the distribution in the figure is indeed directly opposite to that.

Figure 5.7 shows Oort and his family, some colleagues and friends at the dinner after the ceremony.

5.3   Johanna Maria Graadt van Roggen

Although Oort seems to have kept up some correspondence with young ladies he had known in New Haven, and although he will also have had female friends in the Netherlands, there seems to have been absolutely no question of any relationship that could lead to a marriage. That changed when he met Johanna Maria (Mieke) Graadt van Roggen (Fig. 5.8). She was born in Nijmegen on January 7, 1906 and was nineteen when she and Oort met. Mieke's father was Willem Graadt van Roggen (1879–1945) and her mother, Jeanne Laurense Hélène Henriëtte van de Water (1881–1974). Mieke had an older brother Coenraad Jan, who was born in 1904.

Willem Graadt van Roggen had studied law in Groningen, but caught tuberculosis, from which he was slowly recovering for years in his hometown Nijmegen. He wrote poems; in 1903 he published a collection *Het schouwende leven* (the inspective life) and in 1904 a novella *Tubeculeus* on the life of a tuberculosis patient in a sanatorium. Eventually he passed his candidates exam (bachelor) in 1911. He married Hélène van de Water, who was the daughter of a cigar producer Coenraad Jan van de Water (1824–<1916) from Zaltbommel. Although she would have liked to become a singer, she had chosen to be a housewife.

Graadt van Roggen was not very successful as a writer and poet, but without any training he became editor-in-chief of a newspaper in Breda and later in Friesland of the 'Leeuwarder Courant'. He held both positions only for a short time and in 1911 he became editor-in-chief of the 'Utrechtsch Provinciaal en Stedelijk Dagblad', the major newspaper in the city and province of Utrecht and the family moved to Utrecht. As a result of his activities in local politics he became involved in the organization in 1916 of the first *Jaarbeurs* (Annual Fair) in the Netherlands and his newspaper paid a lot of attention to it. It was held in 1917 and 690 companies took part. As a result of this success he became secretary-general of the organization that organized the Jaarbeurs in 1918; he quit his job at the newspaper and started the magazine *De Jaarbeurs*

Fig. 5.8 Picture of Oort and his future wife, probably taken on the occasion of their engagement to marry in February 1926. From the Oort Archives

to promote trade and industry in Utrecht and the surrounding area. He was an adventurous man; in 1917 he made a trip along the Italian and Serbian battlefields (and reported about it in his newspaper), and in 1921 he traveled to Finland when it had just gained independence from Russia. He became a well-known figure, also on the radio, who persuaded companies to come to the Utrecht Jaarbeurs.

Mieke Graadt van Roggen had gone through the HBS for girls and then the 'huishoudschool', the school for housekeeping, which was a school to prepare young girls for the role of house wife or domestic servant. Somehow she had

Fig. 5.9 Oort with his wife and their families after their wedding ceremony. In the front sit Jetske (left) and Emilie Oort. Behind them from the left to the right Engelina Graadt van Roggen-Kan, her husband Coen, father and mother Oort, mother and father Graadt van Roggen, John Oort and his fiancée Pivy van Mesdag. From the Oort Archives

met Oort's brother John, who was studying biology in Utrecht. In his diary Oort wrote on June 23, 1925: 'Meet Mieke Graadt van Roggen'. Since he had returned from New Haven he wrote in English in his pocket diaries. And on June 25: 'Busy day [...] with preparations for the Sempre Ball to which I go with Mieke who is very lovely'. The Sempre Gala is the annual ball of the Leiden Student Association, to which Oort apparently had access. They stayed in touch: in his pocket diaries Oort noted that he received a letter from her and a few times they met together with other friends. For example, at the annual celebration of the 'Leidens Ontzet' (Leiden Relief) on October 3,[2] brother John from Utrecht came with some friends, including Mieke. And on October 26, Oort was in Utrecht, where he went to a theater with, among others, the Graadt van Roggen family. Afterwards he walked with Mieke back to her house

---

[2]This commemorates the breaking of the Spanish Siege of Leiden, on that date in 1574.

in the rain. He slept in his brother's living quarters. Three days later he wrote in his diary (in Dutch, in contrast to his pocket diaries): 'I really believe that I am beginning to fall in love with Mieke. It reflects in my mind and actions which are abnormal'.

Then things moved fast. Mieke Graadt van Roggen stayed at the Oorts in Oegstgeest in December 1925. Oort says about her in his *AIP-Interview* [6] when asked whether she aspired to a career of her own:

> No. She was much interested in poetry and art in general and still is. She never imagined that she would get a career in that line. Her ideal was more to build up a happy household and a harmonious life.

And Oort's younger son Abraham Hans (Bram) Oort:

> She said she found Uncle Jo attractive, but was more impressed with my father. She immediately fell 'in love' with him. I think she was attracted to my father's calm and introverted character. This contrasts with the situation in her parental home with extravert, intelligent parents. In the 'Graadt van Roggen' family in Utrecht there was little peace and quiet and a lot of arguing among themselves, she was not good at it and therefore felt less 'at home'. She still loved her father and older brother, Coen. But there was more disagreement with her mother.

And in another piece:

> She must have been intrigued by my father, Jan Oort, a quiet and self-reliant man. The introvert Oort-family was the opposite of the extravert Graadt van Roggen family. My parental grandparents were teetotalers, while my mother's parents liked a glass of wine.

The Oorts got married on May 24, 1927 in Utrecht (Fig. 5.9). There was a reception afterwards, but there had been one the day before in Oegstgeest. Witnesses were the brothers of the bride and groom, John Oort and Coen Graadt van Roggen. From here on, I will refer to Mieke Oort–Graadt van Roggen as 'Mrs. Oort'.

5.4   The Rotation of the Galaxy

The breakthrough in Oorts research, and the ultimate explanation for the phenomenon of high-velocity stars, came from an article by the Swedish astronomer Bertil Lindblad (1895–1965). In the model he presented the Galaxy as built up of components with increasing flattening and increasing rotation.

The part with the largest flattening and the largest rotation was the disk that we see on the sky as the Milky Way, the least flattened was the system of globular clusters and that rotated little or not at all. Lindblad assumed that the local Kapteyn system was part of a larger system, which had the same center as the system of globular clusters.

Now Oort had been convinced long ago that there had to be absorption of starlight in the Milky Way. In 1926 he was appointed 'privaat-docent' by the university; this was an unpaid job to teach in addition to his appointment at the Sterrewacht. He accepted that appointment with a public lecture entitled *Non-light-emitting matter in the Stellar System*, which was published as a separate publication. Oort also published it in the amateur magazine for astronomy and meteorology *Hemel & Dampkring*, [63]. I have provided an English translation in the Appendix of the 'Legacy Symposium' [11]. In this he concluded that the Milky Way Galaxy had to be much larger than Kapteyn's System and that there had to be significant absorption of starlight.

Inspired by the work of Lindblad, Oort now took two big steps. It was thought (also by Lindblad) that the Kapteyn system was an example of a condensation in the larger system, indicated by Shapley's globular clusters. In the first place, Oort now assumed that it was a continuous system, a large cloud of stars, that extended beyond the center of Shapley's system and that center also had to be the center of its rotation. The boundaries of the Kapteyn Universe in the directions of the Milky Way are then determined by absorption and not by the spatial distribution of stars. And where Lindblad assumed that the rotation would be uniform, Oort took the point of view that the rotation of the cloud of stars should depend on its distance from the center. In that center presumably a large mass was concentrated, which kept the rotation in balance with the gravitational field. In our planetary system, where practically all mass is concentrated in the Sun, the orbital periods of the planets become larger when we go further from the Sun (Kepler's third law). Oort thought it was reasonable that a change in rotational period with distance from the center would apply to the Galaxy also. This property is referred to as 'differential rotation'.

Oort's second important step was his realization that the consequence had to be systematic motions within the cloud of stars. If something turns around uniformly, then everything in it stands still relative to other parts. But in the case of differential rotation this is not the case. Consider Fig. 5.10. For convenience it is assumed that the rotation velocity is the same everywhere in km/s which is not unreasonable at all (but the orbital period then increases with distance from the center). In the middle we have the Sun and a number of stars are

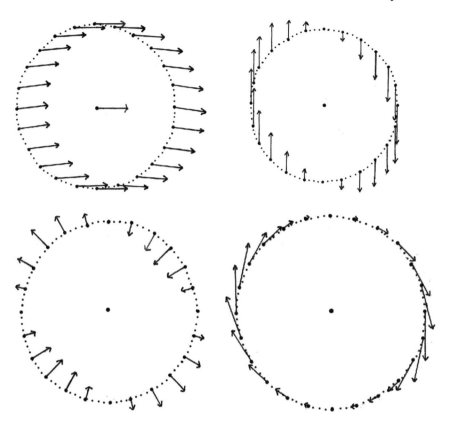

Fig. 5.10 Pattern of motions due to differential rotation. At the top-left we see the Sun in the middle and stars at a distance of 1 kpc from the Sun. All rotations are equal in linear measure (e.g. km/s) and perpendicular to the direction to the center, which is 8 kpc towards the bottom. The figure on the top-right shows the velocities of those stars relative to the Sun. For the sake of clarity, they were blown up by a factor of twenty. The panel bottom-left shows the components of this along the line of sight to the Sun (the radial velocity), and bottom-right that perpendicular to it (the tangential velocities). Figure by the author

shown that are at 1 kpc from the Sun. The arrows then indicate the rotation, with the center lying 8 kpc to the bottom of the page. So the arrows are not parallel and therefore there is motion relative to the Sun. We see this on the top-right, where the velocities have been blown up by a factor of twenty. For each star the velocity relative to the Sun has a component along the line of sight, which we then see at the bottom-left and one perpendicular to it and that is shown at the bottom-right. Now first look at the velocities along the line of sight (the radial velocities). They show a systematic pattern. In the direction towards the center it is zero (because the rotation velocity there is parallel to ours). Then, as we go around counter-clockwise, it grows in the direction away from us and decreases to zero again when we look 90° from the direction to the center. And that pattern repeats itself in each subsequent quadrant. A similar systematic pattern can be seen in the tangential velocities at the bottom right.

Oort collected data to see if such a pattern was indeed present. First, look at the radial velocities. What is expected is a double wave (mathematically speaking a sine). It turned out to be present indeed, but only in a qualitative way due to observational uncertainties. Others had already noticed that there were such systematical effects (including Jan Schilt), but had not appreciated the significance of them. It is understandable that the amount of motion depends on the distance: the further away from the Sun the greater the effect. This is expressed with a constant, which Oort called A. For stars at 1 kpc distance, the maximum radial velocity with respect to us is equal to this constant. The value of constant A was of the order of 10 km/s per kpc (now the best determination is 13), i.e. stars at 1 kpc distance have a maximum radial velocity due to differential rotation of 10 km/s. At 2 kpc it is 20 km/s, etc.

We do not measure the tangential velocity directly, but we determine it through proper motions. This has an advantage, because the amplitude of the pattern then does not depend on the distance. Here we again get a double wave (but shifted by 90°, so mathematically a cosine), but one which is not centered on zero. Oort was able to find this pattern in observations too. That average value (the off-set of the cosine wave) is the second constant and Oort called it B. This one is negative, because the pattern in Fig. 5.10 is clockwise while the Galactic longitude runs counterclockwise; in realistic cases it is of the same order of magnitude as A (the best value now for B is −13 km/s per kpc).

But this pattern also tells us the direction towards the center of the system. After all, in that direction the radial velocity is zero and the tangential velocity is at a maximum 90° away from it. The direction to the center of the Galaxy then turned out to be exactly the same as that to the center of Shapley's system of globular clusters. Everything fitted together. With the Sun we are part of a rotating disk of stars (and dust that absorbs the light), which has the same center as the globular clusters. That system of globular clusters is more or less spherical and will therefore rotate only slowly, as in Lindblad's ideas. This also explains the high-velocity stars: they are part of the same system and their high velocities result from the fact that they have little rotate as a whole; the concentration of the velocities to one hemisphere is the reflection of the rotation of the disk. The reason that it occupies a full hemisphere is that their random motions are of the same order of magnitude as the disk rotation. Now, those velocities must then be directed in the opposite direction as the rotation, and that too was exactly what Oort found. The globular clusters, for which radial velocities had been measured in the meantime, also confirmed this. For pictures of Shapley and Lindblad see Fig. 5.11.

Fig. 5.11 Two astronomers that played important roles in Oort's work on the Milky Way Galaxy. On the left Harlow Shapley and on the right Bertil Lindblad. From the group photo of the General Assembly of the IAU at Harvard in 1932. From the Oort Archives

The average value indicated that the rotation would be 250–300 km/s, if the globular cluster system had no rotation. The rotation velocity is currently believed to be about 220 km/s. The values of the constants A and B, which were soon called the *Oort constants* by everyone, together with the rotation velocity also give an indication of how far the center must be away from the Sun. That turned out to be only 6 kpc, whereas Shapley had estimated the center of his system to be at 15–20 kpc. Oort suspected that the 6 kpc distance would be very imprecise and in the end his intuition turned out to be right.

Oort actually went a step further. He applied the theory of dynamics, developed by British researchers Arthur Eddington (Fig. 5.12) and James Jeans. Galaxy dynamics describes the relationship of the spatial distribution of stars and their motions to the gravitational field. They must be in equilibrium with each other, as long as the system does not contract or expand. The basis, as defined by Jeans and Eddington, consists of two equations. The first one is called the Liouville equation (also called continuity equation), named after the French mathematician and applied physicist Joseph Liouville (1809–1882). The equation says that if there is equilibrium, there must be exactly the same number of stars in each small volume element of the system at any given moment, *and* these must also have exactly the same distribution of velocities. It is also called the (collision-free) Boltzmann equation after Austro-Hungarian physicist Ludwig Eduard Boltzmann (1844–1906). The second equation is the Poisson equation named after the French mathematician Siméon-Denis Poisson (1781–1840), which describes what the gravitational field is if you

Fig. 5.12 John S. Plaskett (left), Arthur S. Eddington and Ejnar Hertzsprung in 1932. This photograph is part of the group picture of the General Assembly of the IAU at Harvard in 1932. From the Oort Archives

have a certain distribution of mass. If one has a given spatial distribution, then one can use those two equations to derive what this means for the rotation and distribution of random velocities in order to have equilibrium.

What Oort did next was use this theory (which he had learned early on from Kapteyn from two fundamental books by Jeans and Eddington) to predict how the stars in the solar neighborhood should move. This was rather mathematical, but in the end Oort wanted to understand the consequences of his findings. An important consequence was that the stars had to have random motions that were on average different in the three fundamental directions. *In* the plane, the mean velocity of stars in the direction away from and towards the center had to be larger than in the direction of and opposite to rotation. And the ratio between these two mean velocities could be expressed in a simple formula that contained only his constants *A* and *B*. Larger velocities from and to the center than in that perpendicular to it give rise to a systematic pattern resembling two opposing streams in the directions with the largest mean motion. This then was the final explanation of Kapteyn's Star Streams.

Oort's work presented a consistent picture of the structure of the Galaxy and the motions within it. Oort published his results in 1927 and 1928. His name was made, when others confirmed his results. Two prominent astronomers, who might otherwise have discovered the rotation of the Milk Way System, are Jan Schilt and especially the Canadian astronomer John Stanley Plaskett (1865–1941), then director of the Dominion Astrophysical Observatory at Saanich, British Columbia (Fig. 5.12). But Oort's fame grew particularly when one of the most prominent astronomers of the time, Arthur Eddington, wrote and spoke about it. Eddington first wrote a widely read review for the Royal

Astronomical Society in London and then chose the rotation of the Galaxy and Oort's work as the subject of his 'Halley Lecture' in 1930. This is a very prestigious lecture organized by the University of Oxford, the first on the occasion of the return of Halley's famous comet in 1910. This comet, named after the British astronomer Edmond Halley (1656–1742), returns to the inner parts of the Solar System about every 76 years.

5.5   Harvard, Columbia or Leiden?

Now that Oort's name had been made, he began to receive job offers from abroad. Not only did people read his papers, he was also present at many large international meetings, where he presented his work. We already saw him as a younger student at the first General Assembly of the International Astronomical Union IAU in Rome in 1922. He also attended the subsequent ones; in fact, he attended all these meetings, that are usually triennial, until the seventeenth of 1979 in Montreal. For a long time he was the only one who had been at all, but now perhaps is still the one who attended more of them than anyone else. After Rome, the IAU met in Cambridge, United Kingdom in 1925 and in Leiden in 1928. At the latter of these, Willem de Sitter was President (Fig. 5.13).

Fig. 5.13 The executive board of the IAU on the steps of the Sterrewacht Leiden during the General Assembly there in Leiden in 1928. In the front on the left Henri-Alexandre Deslandres (1853–1948) from Paris, President Willem de Sitter, Shin Hirayama (1867–1945) from Japan. In the back Frank Schlesinger, General Secretary Frederick John Marrian Stratton (1881–1960) and Arthur Eddington. From the archives of the Sterrewacht Leiden

The first major offer came shortly after this meeting in Leiden in the form of an extensive letter from the director of Harvard College Observatory, Harlow Shapley. The letter was dated December 22, 1928. It did not concern an appointment at that observatory, but a professorship at the prominent Harvard University in Cambridge (Massachusetts) near Boston. Oort could use the facilities of Harvard College Observatory, including telescopes in the southern hemisphere. Shapley suggested a possible sabbatical year at a planned 60 in. (1.5 m) reflector in South-Africa. This telescope was eventually put into operation in 1933 at Harvard's observing station (Boyden Station) in Bloemfontein. The southern station that Harvard already had in Arequipa, Peru, with the 24 in. (0.6 m) Bruce telescope, which Pickering had used for the southern *Selected Areas* of Kapteyn, was also moved to Bloemfontein in 1929. Oort was especially interested in this facility to study the motions of stars and he hoped to get an important part of the observation time.

Oort asked advice from various colleagues, including Willem de Sitter and Ejnar Hertzsprung in Leiden and his thesis supervisor Pieter van Rhijn in Groningen. But also Frank Schlesinger, whose advice he clearly took most seriously. The latter pointed out that Oort would have many more possibilities in America; in Leiden he might succeed de Sitter as director, but that would take another thirteen years. On the other hand, it concerned a position with quite a heavy teaching load and Schlesinger warned that the informal link with the Observatory could be terminated under different management. But Oort could, of course, take another job in the United States after a few years. After careful consideration, Oort decided not to accept the offer, mainly because of the heavy teaching load and the uncertain relationship with Harvard College Observatory.

The second offer came not long after that. Schlesinger wrote Oort on January 25, 1930 that Oort would be visited in Leiden by a professor from Columbia University in New York city (Upper Manhattan)—one Prof. Barry—who would offer him a job at that university. This seemed like a much better option. Columbia was a leading university and Schlesinger felt that this presented an excellent opportunity. Oort reported to Schlesinger that his warning letter had arrived two days late, Prof. Barry had already come unannounced. Oort had two objections, as he wrote to Schlesinger. The first was Mrs. Oort, who was 'very attached to Holland and the people here and she absolutely dislikes the idea of having to live in another country'. And Oort himself had reservations about the fact that he had understood from Prof. Barry that the offer was about setting up a theoretical department. But Oort wanted to keep in touch with observational astronomy. He would consider the offer if Columbia was prepared to support a project for a large (he meant 60 or 70 in., say 1.5–1.75 m) telescope in the southern hemisphere.

Schlesinger objected that he did not consider it a good idea for Columbia to build a southern telescope in view of the many existing plans of other universities to do so. He warned Oort that if he rejected this second special offer, it might be that he would receive no more attractive offers. In any case, Oort made his wishes known to Columbia that in order to accept he had to have an assurance that funds would be available for the construction of a southern telescope.

On June 29, 1931 Oort wrote to Schlesinger: 'In the end we have decided to stay in Holland'. And Mrs. Oort wrote to her parents:

> Here's a delightful piece of news for you: We are not going to America! The answer came and was 'no telescope', so it's decided now. Lovely, isn't it.
>
> It is a bit of a disappointment for Jan, of course, because of the possibility of a beautiful piece of work, but there is a lot against it, and we are looking forward to a future in Holland.

Indeed, no more special offers came from the United States after this. Over the years, Oort has spent several periods of several months at major American universities and observatories on invitation and lectured there, but long-term association or emigration was now definitely out of the question. Incidentally, in 1943 Columbia decided to participate in the 26 in. Yale telescope of Schlesinger in South-Africa. And for the sake of completeness: Columbia University appointed Jan Schilt as director of the astronomy department in 1933, where he remained until his retirement in 1962.

Fig. 6.0 Photograph of the Oorts in 1926. From the collection of Abraham H. Oort

6

The Structure of Our and Other Galaxies

*In a sense, the galaxy hardest for us to see is our own. For one thing,
we are imprisoned within it, while the others can be viewed
as a whole from outside. Furthermore, we are far out from the center,
and to make matters worse, we lie in a spiral arm clogged with dust.*
Isaac Asimov [64].

*'Then how do you know it's there?' He slows to a stop at a red light.
'Dark matter has a gravitational effect on other objects. You can't see it,
you can't feel it, but you can watch something being pulled in its direction.'*
Jodi Lynn Picoult, (b. 1966) [65].

After their marriage the Oorts lived for some time at the address Plantsoen
35 in Leiden. But soon they moved to Cobetstraat 63, probably anticipating
upcoming family expansion. There their eldest child, Coenraad Jan (Coen),
named after Mrs. Oort's brother, was born on December 5, 1928 (see Fig. 6.1).
Both addresses were a short walk from the Sterrewacht. The other two children
came some years later, daughter Marijke on April 25, 1931 and younger son
Abraham Hans (Bram) on September 2, 1934. The daughter was not named
after family members; the Oorts apparently thought Marijke was a beautiful
name. Son Bram was named after Oort's father Abraham and mother Ruth
Hannah.

But there was also sadness. A big blow to the family was the sudden death in
August 1933 of Mrs. Oort's brother, Coen Graadt van Roggen, at the age of only
29 years. He had studied law in Utrecht and had also obtained a PhD there.

Jodi Picoult is an American writer. This quote is from her novel *My sister's keeper*.

P. C. van der Kruit, *Master of Galactic Astronomy: A Biography of Jan Hendrik Oort*,
Springer Biographies, https://doi.org/10.1007/978-3-030-55548-1_6

Fig. 6.1 1 This photo dates from 1929, when the first child of the Oorts, Coenraad Jan, was only a few months old. We see Mrs. Oort in the middle with her in-laws and sister-in-law Jetske Oort. From the Oort Archives

He worked as a promising editor at the newspaper 'Nieuwe Rotterdamsche Courant', where he specialized in movies reviews. He was one of the first to consider movies as a form of art. Apart from his wife, he left behind two children.

6.1   The Oort Limit

Oort continued his analyses of the distribution of stars in the Milky Way Galaxy and their velocities and of the dynamics of the Galaxy with several extensive studies, which sometimes took years to complete. His great mentor Kapteyn had presented a first analysis of the dynamics of the Galaxy in his famous paper of 1922. As we saw, part of it was that the explanation of the 'horizonal' structure involved a opposite rotation of two components of stars around a center at some 650 pc distance, which manifested itself as his two Star Streams. Mainly because of Oort's work this was now replaced by differential rotation around a much more distant center that is hidden from our view by interstellar dust. In the new understanding the streams are in fact manifestations of an asymmetry in the random velocities of the stars in the solar neighborhoods, which on average are greatest in the direction toward and opposite to that center. Kapteyn's first part of the study had concerned the vertical distribution and dynamics of his system, and Oort took up this work also. It had become a study of the structure and dynamics of the disk of the Galaxy in the direction perpendicular to the plane.

The method was in fact the same as that of Kapteyn, but Oort was in a position to investigate many more details. Kapteyn had simply taken the vertical distribution of the stars in his model and assumed that this also reflected the full distribution of matter. Oort's analysis was more sophisticated. He concentrated on groups of stars, of which he then determined the vertical distribution separately. He then also determined the velocities of these stars in the vertical direction. If the system is in equilibrium, one can consider those stars as test particles whose spatial distribution and velocities can be used to determine the gravitational force as a function of distance above the plane. This in turn could be used to determine the distribution of all matter.

Oort distinguished various spectral types of stars. This means that he made a distinction between the sorts of stars that together formed the population of stars in the solar neighborhood. A brief summary of the structure and evolution of stars, which is necessary to understand what was meant, follows here; a somewhat more comprehensive treatment can be found in Appendix A. On the basis of the presence of absorption lines in spectra of stars they are divided up in a series marked with the letters $OBAFGKM$. The majority of the stars lie on a diagonal line in the fundamental diagram of stars, the Hertzsprung–Russell diagram. This diagram shows the relationship between the luminosity of a star and its surface temperature. The first follows from the apparent magnitude in the sky and the distance; the second follows from either the color of the starlight or the spectral type. Almost all stars (among which the Sun which is of spectral type G) lie on a diagonal line, the Main Sequence, in the Hertzsprung–Russell diagram and 'burn' hydrogen into helium in their central parts. The most massive stars are hot and bright (spectral type O) and take the shortest time to convert the available hydrogen in the center into helium. The Sun takes ten billion years for that and is about halfway now. The lightest stars (spectral type M) are faint and red and use their hydrogen much more economically; they use it up on a timescale of ten times the current age of the Universe. Stars, that have passed that Main Sequence phase, are mostly red giant's, that are bright but cool (spectral types K and M) and are found to the upper-left in the diagram. This phase does not take very long. Most of the stars we see in that phase are a bit more massive than the Sun and about the same age as the oldest stars. Because they are intrinsically bright (and easy to recognize) and can be observed from a great distance, Oort began his analysis with those stars.

From various places in literature, Oort collected everything he could find out about *K*- and *M*-giants that are far from the Milky Way in the sky, so that they are also spatially outside the Milky Way plane. Oort took those stars for which the absolute magnitude had been determined and the radial velocity had been measured (because these stars are seen more or less perpendicular from the plane this can be converted into the vertical velocity with a small correction). He then derived the density of such stars as a function of distance from the plane of the Milky Way. Together with the distribution of their velocities, which appeared independent of distance from the plane, he calculated the vertical force needed to maintain an equilibrium between those velocities and their spatial distribution. In this way he could calculate the vertical force to a distance of about 500 pc from the plane.

Oort next turned to other spectral types. For different types he then found that they had different average velocities, the *O* stars the smallest. What Oort did then was to extrapolate the vertical force to greater heights and calculate what the average velocities for various groups of stars had to be in order the give rise to the observed spatial distribution. And when he had done that for all the such groups of stars, he could use that extrapolation to predict what the number of stars in the sky should be as a function of of apparent magnitude, which could then be compared to the observations and select the correct extrapolation of the force. That required an enormous amount of numerical work performed by computers.

Eventually, Oort followed this same procedure for directions at different angles with the Galactic plane, that are not perpendicular to it, except for directions so close to the Milky Way that absorption would be a problem. And this way he found what the distribution of stars in the Galaxy should be. He presented this spatial representation as a cross-section of the Galaxy, which is reproduced in Fig. 6.2.

Oort went further still. Now that he knew the vertical gravitational force, he could use it to calculate the total distribution of matter to produce that force. That can be characterized by the amount of total mass density in the Galaxy near the Sun and that amounted to 0.092 solar masses per cubic parsec ($M_\odot/pc^3$). The known visible stars give a total of 0.038 $M_\odot/pc^3$. But the lightest stars have a mass 0.22 $M_\odot$ and it was known that there were very many lighter object, which would be intrinsically much fainter or not emit much light at all. Oort estimated that these could contain enough matter to explain the value for the total mass density, so that there was nor need to invoke large amounts of dark matter (dust or other 'meteor-like' material). To this day, this has still not been conclusively resolved. Oort's 0.092 $M_\odot/pc^3$ was soon called the Oort Limit.

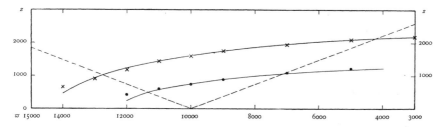

Fig. 6.2 Cross section through the Galaxy, as calculated by Oort in 1932. Symmetry is assumed with respect to the plane of the Milky Way; the Sun is at 10,000 pc along the horizontal axis and the center of the system is to the right at zero, which is beyond the extent of the figure. The curved lines represent fits with ellipsoids centered on the center and follow equal 'density of starlight', the total light emitted by the stars in a small volume. For the lower curve this is 1/25 of what it is near the Sun and for the upper one it is 1/100. The dotted lines show latitude 20°; below this latitude the analysis is uncertain due to interstellar absorption. From the Oort Archives

The amount of calculation necessary for this analysis was enormous. The Sterrewacht employed 'calculators', who did not much else but perform calculations by hand (in principle everything was done twice for verification purposes). The importance of these people should not be underestimated. Figure 6.3 shows the staff of the Sterrewacht in 1931. Among them are ten such calculators. This picture shows the senior astronomers in the front row, before the steps to the Sterrewacht entrance. The two calculators who worked for Oort on the work just discussed are Gerrit Pels (1893–1966), third from the right, and Johannes Marinus Kriest (1895–1981), at the back, straight above Hertzsprung. The third person from the left is C. Sanders, who was involved in measuring absolute declinations from Angola (Sect. 4.1).

6.2   Perkins Observatory

At a very early stage Oort had become interested in other galaxies, usually referred to as extragalactic. The first question was whether the Kapteyn system, to which we belong, had the same structure. With the new insights it became clear that the larger Galaxy was one of many other galaxies, but it was especially interesting to answer the question how our Galaxy would fit into the diversity and especially what the dynamical structure of other galaxies was. Oort had met Edwin Hubble during his visit to Mount Wilson Observatory in 1924. There he had been very impressed by Hubble's beautiful pictures of galaxies. In the Oort Archives one can find many sheets of paper with annotations about individual galaxies and their parameters.

There is a great diversity among galaxies, which is described in the classification system that Hubble defined in the 1920s. It is usually not mentioned

1. G. van Herk
2. P. Th. Oosterhoff
3. C. H. L. Sanders
4. Ir. W. E. Kruyt-
    bosch
5. J. Uitterdijk
6. Dr. C. H. Hins
7. A. J. Wesselink
8. Miss H. A. Kluyver
9. G. P. Kuiper
10. Dr. J. H. Oort
11. A. de Sitter
12. Miss C. H. de Nie
13. H. Kleibrink

14. P. P. Bruna
15. Prof. Dr. E. Hertz-
    sprung
16. J. M. Kriest
17. M. D. Schepper
18. B. C. Mekking
19. L. Gaykema
20. J. C. Gaykema
21. P. de Haan
22. J. E. Prins
23. H. Zunderman
24. Prof. Dr. W. de
    Sitter
25. J. E. Stol

26. L. J. F. van Leeuwen
27. Miss W. van Heuzen
28. A. Luteyn
29. D. Gaykema
30. J. H. Kasten
31. H. M. Swaak
32. Dr. J. Woltjer
33. J. Nicaise
34. G. Pels
35. F. de Haas
36. E. W. de Rooy

(Mr. C. J. Kooreman was
absent on military service)

Fig. 6.3 The staff of the Sterrewacht Leiden in 1931. This picture was published by W. de Sitter in a book describing 300 years of history of the Sterrewacht [66]. In the front row from left to right Coert Hins, who effectively led the astrometric department, Oort, Hertzsprung, de Sitter and Jan Woltjer. From the Oort Archives

Fig. 6.4 Diversity among galaxies. From top left to bottom right the systems M87 (an elliptical system), M81 (an Sa spiral galaxy), the Andromeda Nebula M31 (an Sb spiral) and M101 (an Sc spiral). Except M31 these pictures were obtained with the Hubble Space Telescope. Credit: NASA, ESA, the Hubble Heritage Team, Digitized Sky Survey 2 (STScI/AURA) [67]

that Hubble based his work largely on that of British astronomer John Henry Reynolds (1874–1949), which Hubble failed to acknowledge.[1] Figure 6.4 shows the most important aspects on the basis of some bright galaxies from the catalogue of 110 bright nebulous objects compiled by Charles Messier (1730–1817) and published in 1771. At the top-left an elliptical system without a disk and to the right and the bottom an increasing contribution of a disk with increasingly better defined spiral structure. Figure 6.5 shows two systems edge-on. The lower one (NGC891) is more or less a twin of our own Galaxy, as I have documented on several occasions (for example in a conference about Kapteyn [11]). We already saw that our Galaxy consists of a halo that is more or less spherical and does not rotate much (and contains, among other things,

---

[1] For more background and discussion on the issue of Hubble's discoveries and his systematic failure to give the proper credit to others that they deserved, see the proceedings of a conference on *Origins of the Expanding Universe: 1912–1932* [68].

Fig. 6.5 Two galaxies seen edge-on. Above M104, also known as the Sombrero Nebula, and below NGC 891. Because the latter galaxy has many parameters that are similar to those of our own Milky Way Galaxy, I regard it as our 'twin galaxy'. Credit: NASA and The Hubble Heritage Team (STScI/AURA) and NOAO [70]

globular clusters and old stars), and a rotating disk, to which the Sun belongs. The sequence of galaxies is one from a dominating halo to a dominating disk. For more details on structure and dynamics of galaxies see e.g. my course on this subject [69].

Oort's interest in extragalactic systems concerned the study of their structure and dynamics. Of course he also wanted to know if our Galaxy had spiral structure and if so, what it looked like. But there was no chance to find out much about this, because the interstellar dust only allows us to study the immediate surroundings of the Sun. Dynamics of other galaxies was observationally not easy at all. You had to determine the three-dimensional distribution of stars and also the internal velocities of the stars in it. To complicate matters further, these galaxies themselves might also contain absorbing dust. The first step was to measure the distribution of brightness across the image of a system. This is called surface photometry and not much of it had been done yet. The procedure is to take a photographic plate and then measure point by point in the system how dark the emulsion has become. Oort began by attempting this first step.

But how would he go about obtaining the necessary observational material? That required a large telescope, like the ones at Mount Wilson in California. But to go there for photographing galaxies was well beyond his financial means. Now, in the early 1930s, a new telescope was built near Delaware, Ohio, and that would present a possibility. It was the Perkins Telescope.

The Perkins Observatory was associated with Ohio Wesleyan University, a so-called liberal arts college dedicated to a broad interdisciplinary education (literature, social and natural sciences). It was founded in 1842 by followers of John (1703–1791) and Charles (1707–1788) Wesley, methodists who wanted to revive the Church of England. The university is located in Delaware, Ohio, about 45 km north of Columbus, the capital of Ohio. Hiram Mills Perkins (1833–1924), a very religious man, taught mathematics and astronomy there, for which he set up an observatory for students in 1896. He had made his fortune selling pigs during the Civil War to feed the soldiers; the pigs were bred on his family's farm. After his retirement in 1907, he designated a substantial part of his fortune to the construction of a large telescope in Ohio, which should be the largest in the United States after the 100 in. of Mount Wilson. The place that was chosen to locate it was just outside Delaware. Eventually it became a 69 in. (1.75 m) reflector (mirror-)telescope (Fig. 6.6). Of course, he climate of Ohio is totally unsuitable for astronomical observing and the telescope was eventually moved to Lowell Observatory near Flagstaff, Arizona in 1961.

In April 1932 Oort approached the director Harlan True Stetson (1885–1964) (see Fig. 6.7). At that time the mirror had just been placed in the telescope. Oort suggested that he would come to Ohio for a few months to use the new telescope to take plates of galaxies for surface photometry. He was prepared to give advanced lectures in return (e.g. on statistical astronomy, the work of Kapteyn). Stetson approved of the idea and arranged for Oort to lecture to PhD students at Ohio State University in Columbus. Stetson and Oort had apparently met earlier in New Haven, where Schlesinger every

Fig. 6.6 The 69 in. (1.75 m) telescope from Perkins Observatory. This photograph comes from an old postcard, which refers to the fact that this was the third in size in the world, after the 100 in. of Mount Wilson and the 72 in. (1.83 m) of the Dominion Astrophysical Observatory in British Columbia, Canada. [71]

month brought astronomers from the general area together. Before coming to Perkins Observatory Stetson had taught astronomy at Harvard (but not to advanced students; the latter was part of the position that had been offered to Oort earlier).

Oort's plan was to combine this with a trip to the US to attend the IAU General Assembly that year at Harvard University in Cambridge, Massachusetts. Mrs. Oort traveled with him. The (then still two) children Coen and Marijke were three and one years old and stayed with their grandparents in Utrecht. The Oorts left Rotterdam on August 12, 1932 with exactly the same boat as with which Oort had traveled to New Haven in 1922 (but now first class). In New York they stayed for a few days with Jan Schilt and his wife Jo, before traveling on to Randolph, New Hampshire, where the Torreys had moved after Charles' retirement. From there they went to Magog in Quebec, just across the Canadian border to assist Utrecht astronomer Marcel Gilles Jozef Minnaert (1893–1970) in observing a total solar eclipse. Minnaert was Flemish; he had had to flee in 1918 because he had collaborated with the German occupation in order to establish a Dutch-speaking university in Ghent. Minnaert was actually a biologist by training, but eventually became an important astronomer/solar physicist. The solar eclipse project was a complete failure due to clouds. Next the Oorts travelled to Boston and the IAU. Oort played a modest role there.

Fig. 6.7 Oort and Mrs. Oort in the group photo taken at the IAU General Assembly in Cambridge, Massachusetts, in 1932. The person in front of Oort (wearing glasses) is Harlan Stetson, director of Perkins Observatory, and on the right his wife. From the Oort Archives

Before leaving Cambridge they bought Harlow Shapley's car, a 1925 Buick Sedan (see Fig. 6.8 for a sample of that type). With this car, they first travelled to New Haven to visit the Schlesingers. It was Mrs. Oort's first trip to America, so it was also an opportunity to introduce her to the places where Oort had worked and the people he had met there. In New Haven they also saw Helena Aletta (Heleen) Kluyver (1909–2001), who had studied astronomy in Leiden and still worked at the Sterrewachr. She had also visited the IAU (see Fig. 6.10). The Schilts also came to New Haven and together they did quite a bit of sightseeing. The Oorts then travelled on to Delaware, where they arrived on October 4. Even travelling by car was slow by modern standards. They drove via Pittsburgh and Oort noted that they covered the remaining 146 mi (235 km) from Pittsburgh in six and a half hours; this means an average speed of 36 km/h.

The work at the telescope (Fig. 6.9) encountered a lot of setbacks, mainly due to the weather that was unsuitable for astronomical observations anyway in Ohio. In addition, there were problems with the shutter and astigmatism (when the mirror does not focus all the light of a star into a very small spot). So the material that Oort collected was minimal. According to a letter from Mrs. Oort to her parents, it was not a complete failure at all however, because Oort had talked to many astronomers, had gained new ideas and felt scientifically refreshed. During November they had made a trip with their car to Ann Harbor (Michigan) and Chicago, where Oort visited colleagues. On their way back to

Fig. 6.8 Picture of a 1925 Buick Sedan, the type of car that the Oorts had bought from Shapley and with which they travelled during their 1932 trip to Perkins Observatory in Delaware, Ohio, and other places [72]

Fig. 6.9 The Oorts at the Perkins 69 in. Telescope in an article in the Columbus Dispatch. Mrs. Oort poses as a assistant with the observing; however, there was a PhD student working under Stetson who in reality did the work. From the Oort Archives

New York they visited the Niagara Falls, but these were frozen. On December 20, they left New York and were back in the Netherlands just before New Year. There is no record of how they disposed of the car.

The technique for photographic surface photometry is to expose, in addition to an image of a galaxy, on a part of the plate (or a separate plate) an image of an area of stars of known magnitudes (e.g. in one of Kapteyn's *Selected Areas*). Of course stars would easily overexpose the plate, so for that exposure the telescope had to be taken out of focus, so that the light was dispersed over a larger area and stars produced blurred images. Later that could then be used to relate the darkening of the emulsion to the amount of light exposed.

Measuring the plates that Oort had obtained required a special instrument. Fortunately, Minnaert had developed such an instrument in Utrecht to measure

Fig. 6.10 Some persons important to Oort from the group photo of the staff of the Sterrewacht in 1931, see Fig. 6.3. On the left Heleen Kluyver, to her right Dirk Brouwer, who was to succeed Schlesinger as director of Yale Observatory, in the middle Aernout de Sitter with whom Oort had developed a very good friendship, and on the right calculator, later photographer Herman Kleibrink

photographic recordings of the Sun's spectrum for his large project of producing a Solar Atlas. Oort was allowed to use this for his surface photometry. In practice the work was done by one of the calculators, in this case Herman Kleibrink (1910–1991), see Fig. 6.10. The results were disappointing; stellar images were sometimes shifted as much as 10 mm during the exposure. After some correspondence with Stetson it was concluded that the primary mirror could shift during the changing orientation of the telescope while the object was followed in the sky. In the end, the project with the Perkins telescope was declared a failure.

Oort had another trick up his sleeves. Pieter Theodorus Oosterhoff (1904–1978), after completing his PhD with Hertzsprung in 1933 on a subject concerning photometry of stars, worked for some time at Mount Wilson Observatory. With the permission of the director Walter Adams, he obtained a small number of plates for Oort on the 60 in. Telescope and that turned out to be successful, as we will see later.

6.3   The Succession of de Sitter

Willem de Sitter died quite suddenly on November 20, 1934, after a short illness. Probably this had to do with the tuberculosis that he had contracted after an overdose of ether for anaesthesia during surgery. For this he had spent a period of recovery between 1919 and 1921 in a sanatorium in Arosa, Switzerland. The question now was who should succeed him (see also [73]).

De Sitter was 62 years old and he would have retired only in 1942. The expectation had been that Oort would be the most likely successor of de Sitter. In that year Hertzsprung would be 69 and only one year away from his retirement. The situation now was very different, Hertzsprung was 61 years of age and Oort 34. Hertzsprung had been appointed as a extraordinary professor, i.e. his chair was not part of the regular contingent of professorships and it also paid less. In addition, Hertzsprung was appointed deputy director of the Sterrewacht. He himself felt—understandably—that he should take over the positions of professor and director that de Sitter had held (see Fig. 6.11). There was a problem with that, because Hertzsprung was known as a workaholic and a not very social person. The concern among the staff was that their working conditions would become less pleasant under a probably very authoritarian director Hertzsprung. For the same reason, there was some reluctance within other parts of the university.

For Oort, the primary consideration was that he was not sure that he would have the same freedom and support for his scientific research under a Hertzsprung directorate as he had enjoyed up till then under de Sitter. Part of the issue was access to the Leiden southern station. On his appointment Hertzsprung had been promised access to a telescope in the southern hemisphere to perform photometry of stars. This was first in the form of an agreement with the Unie Sterrewag (Union Observatory) in Johannesburg. Hertzsprung also had access there to the 10 in. Franklin-Adams Telescope, named after John Franklin-Adams (1843–1912), who had donated the telescope. The telescope had been the first to produce a photographic atlas of the entire sky (the northern sky had been covered first with the same telescope at Franklin-Adams' private observatory in England). Hertzsprung used it to study variable stars by taking repeated exposures. In 1930, the Rockefeller Foundation had made funds available for the construction of a telescope in the south for the Sterrewacht Leiden, for which de Sitter had supplied the specifications. The grant was administered in a foundation called the Sterrewacht Fund, from which the operations of the Leiden Southern station were financially coordinated. So there was a southern station of the Sterrewacht Leiden on the site of the Unie Sterrewag. The cooperation was excellent; Willem Hendrik van den Bos (1896–1974), a student of de Sitter, had been the Leiden representative there since 1925 until he became staff member of the Unie Sterrewag itself in 1928 and Hendrik van Gent (1900–1947) took over the Leiden position. Van Gent had obtained his PhD with Hertzsprung in 1932. The Rockefeller Astrograph, built with the grant, was a double astrograph, especially designed for photographic studies of the sky. Oort now did not want to lose the op-

Fig. 6.11 Willem de Sitter, director, and Ejnar Hertzsprung, adjunct director, around 1930, probably at the Sterrewacht Leiden. From the Oort Archives

portunity to gain access to this facility, which would not have been a problem under de Sitter, but under Hertzsprung he was not quite sure about that.

A period of tug-of-war, postponement of decisions and uncertainty followed. Various solutions were proposed whereby the positions of ordinary and extraordinary professor and of director and deputy were divided between Oort and Hertzsprung in various ways. Hertzsprung, however, wanted absolutely nothing less than both senior appointments for himself. The university could not arrive at a decision and the Minister himself had to resolve the issue. There is a note from Oort with discussion points in preparation for his meeting with the Minister, in which he indicated that he would give in to Hertzsprung if he were given sufficient guarantees of independence. After some hesitation he chose to wear a dress suit. In the end Hertzsprung became professor and director and Oort extraordinary professor and deputy director. This had been agreed between Hertzsprung and Oort, especially stressing the point that

Fig. 6.12 Sterrewacht Leiden in 1933 as seen from the northwest (compare Fig. 5.1). On top of the main building we see the dome of the 10 in. refractor (this is a telescope with a lens rather than a mirror as primary optics), and that of the 6 in. refractor just visible behind the smaller dome. On the right part of the housing of the meridian circle. Next to and beyond the main building were living quarters then used by Hins and Hertzsprung. In the background the building of the photographic department and the dome of the photographic telescope. Sterrewacht Leiden Archives

Oort remained responsible himself for conducting and chosing topics for his research.

A second issue had been added (apparently a bit late in the proceedings). The senior staff of the Sterrewacht made use of houses on the premises. The Oorts now had three children and living in a house on the Sterrewacht would be very convenient. Now the most attractive home was that of the director, where de Sitter had lived until his death. Hertzsprung lived in a smaller house (see Fig. 6.12), and was of course entitled to move to the director's house. But Oort felt that house was much better suited for him and his family. Hertzsprung lived by himself (he had been separated from Kapteyn's daughter Henriette since 1923). As part of the final arrangement, Hertzsprung refrained from exercizing his right to move into the director's house, which probably must not have been a difficult decision or concession for him.

The Oort family moved into the director's house of the Sterrewacht with its spacious garden (see Figs. 6.13 and 6.14). Oort occupied a large study adjacent

Fig. 6.13 The front of the director's residence on the Sterrewacht, photographed from the Sterrewachtlaan. It lies to the right of the view in Fig. 6.12. The person on the portal is the youngest son Abraham (Bram) Oort. From the collection of Abraham H. Oort

to the house, on the right-hand side of the landing in Fig. 6.14, which became his office at the Sterrewacht.

Oort accepted his appointment with an inaugural lecture, *The construction of the Stellar System*. He published it in the amateur magazine for astronomy and meteorology, *Hemel & Dampkring* [74].[2] He began this lecture with a description of the very irregular distribution of bright galaxies across the sky. There is a strong concentration, which is above the horizon in the middle of the night in spring as seen from the northern hemisphere (around March/April). This is a large collection of galaxies called the Virgo Cluster (in the constellation of Virgo) at about 16 million parsec (Megaparsec or Mpc), corresponding to 50 million light-years, away from us. For galaxies at fainter magnitudes, the distribution became more and more regular, as Hubble had informed him. Remarkable also was what had long been referred to as the Zone of Avoidance around the Milky Way. Due to the absorption of dust in the Milky Way plane, we cannot see any extragalactic systems there. Oort described the results of his research into the structure and dynamics of the Milky Way Galaxy up till then, and paid relatively much attention to the question of what our Galactic System would look like when seen from a great distance, speculating that it would probably have spiral structure. That was an issue that would occupy him for years to come. The dynamical properties that were known, indicated that

---

[2]For an English translation see Appendix B3 of [2].

Fig. 6.14 The residence of the director of the Sterrewacht seen from the garden side (Witte Singel). Daughter Marijke Oort is on the balustrade. From the collection of Abraham H. Oort

such a structure would not withstand the effects of the differential rotation for very long. Remember, that at that time it was not even clear whether that rotation was in the direction of winding up or unwinding the spiral arms (it turned out to be the first). What process maintained the spiral structure?

6.4   An Expedition to Kenya

At the time of de Sitter's death, the Rockefeller Astrograph for the Leiden southern station was still under construction. Although Hertzsprung regarded it primarily as his telescope, we have seen that Oort was also interested in it. This is also clear from the fact that in 1938, when the telescope was put in operation, he wrote an article about it in *Hemel & Dampkring*. However, Oort never has made much use of it. In 1931, Aernout de Sitter had taken up the position of Leiden observer in Johannesburg. He had written an PhD thesis under Hertzsprung. The Oorts had become very good friends of de Sitter and his wife and, although he was a man of limited capabilities Oort had done his best to secure a good job for him.

After the publication together with Oort on absolute declinations, and of the work of Oort and Sanders on the measurements of these near the equator, Willem de Sitter had made preparations to carry out this program and received financial support for this from the IAU. This led eventually to the first of what

Fig. 6.15 The observation station at Equator in Kenya during the first expedition. The roof of the hut is opened up completely during the observing. In the middle a zenith telescope to precisely determine the actual geographical latitude. On the right the horizontal telescope for measuring azimuths. Sterrewacht Leiden Archives

would become two expeditions to Kenya to establish a system of absolute declinations from the equator (see [75] for an extensive description and discussion of those expeditions). Staff member Coert Hendrik Hins (1890–1951) (see Fig. 6.3) led the expedition and was accompanied by young astronomer Gijsbert van Herk (1907–1999), (far left in Fig. 6.3). They left in July 1931 and built a temporary observing station in a place they had chosen after a carefull search near a railway station aptly named Equator (Fig. 6.15). It was far enough from Mount Kenya to have a relatively clear horizon and was 300 km from Lake Victoria, which by evaporation could create fog; however, the latter proved to be a serious problem anyhow. Despite the position of almost exactly on the equator it was very cold and windy at the altitude of 2850 m during the observations (it froze regularly) and the work of Hins and van Herk took place under extremely harsh conditions and was accompanied by many hardships. Because they had to make observations close to the horizon, there was also no protection against the sometimes strong wind. There were also leopards and wild boars. The observations lasted from December 1931 to February 1933 for a total of 453 nights, of which only 123 proved useful. Hins and van Herk stayed all that time in Kenya, although occasionally one of the two spent a few days at lower altitudes. They returned to Leiden in May 1933.

Part of the programme was intended to become van Herk's PhD thesis to defend with de Sitter as his supervisor, but after the latter's death Oort took over that function. The defence took place on May 1, 1936. Already during the measurements in Kenya there had been problems with inconsistencies in the observations. After a lot of reduction work, van Herk and Hins came to the conclusion that the accuracy of the declinations was no better than a disappointing $\pm 0''.92$, while the expectation had been that it would be $0''.35$. This appeared to be the result of hysteresis in the horizontal telescope. Each instrument has errors in collimation, i.e. deviations in the alignment of the optical components. To correct for this, the instrument is rotated regularly around its long axis so that the errors go in the opposite direction. This way corrections can be determined and applied. Hysteresis means that there are shifts in the optics, depending on which way you turn the telescope around its long axis. When that is the case, the alignment errors cannot be determined properly. It also turned out that the length of the period of observation in Kenya had been far too short. Oort therefore decided to organize a second expedition. However, the Second World War prevented the implementation of this plan until the end of the 1940s.

6.5   General Secretary of the IAU

The General Assembly of the International Astronomical Union of 1935 was held in Paris. Frank Schlesinger was president. Unlike the President, who was (and still is) appointed for three years, at that time the General Secretary, who did a lot of the practical work, remained in office as long as he or others wished. British Frederick Stratton had held that position since 1925, but now wanted to resign. Oort was asked to succeed him; Schlesinger must have had a hand in this. It was a relatively time-consuming job, but in practise Heleen Kluyver did a large part of the routine work. She worked as a secretary at the Sterrewacht, but since she had actually studied astronomy, she was also involved in scientific research. In 1954 she finally obtained a PhD on the basis of a thesis on lines in spectra of certain pulsating stars; her supervisor was Pieter Oosterhoff, who had in the meantime become a professor in Leiden.

The first General Assembly with Oort as General Secretary was held in Stockholm in 1938. Sweden was the homeland of Bertil Lindblad. Mrs. Oort travelled with him and they stayed with the Lindblads for some time (see Fig. 6.16). The threat of the Second World War was very real, as the new President, Arthur Eddington, emphasized in his closing remarks. Oort was reconfirmed as General Secretary, and the next General Assemble was planned for 1941 in Zürich, Switzerland.

Fig. 6.16 Mrs. Oort with Bertil Lindblad and his wife, Dagmar Bolin (1896–1986), during the visit of the Oorts to Sweden in connection with the IAU General Assembly in Stockholm in 1938. This picture must have been taken by Oort. From the Oort Archives

6.6   Absorption and the Structure of the Galaxy

In his *Plan of Selected Areas* Kapteyn had envisioned that in his 206 areas in the long run star counts would be made and parameters would be determined from as many stars as possible. Then a comprehensive analysis of the spatial structure of the Stellar System would be possible. I already mentioned that Harvard Observatory and its southern station had taken photographic plates of all those areas (and at the insistence of Harvard director Pickering—and frustration of Kapteyn, who thought it was too much work—even in an extra number in the Milky Way). After Kapteyn's death the work was continued in Groningen by Pieter van Rhijn. The joint publications between Groningen and Harvard were concluded with a final volume in 1930. In all *Selected Areas*, stars were counted to magnitude 16, which is about 10,000 times fainter than the naked eye can see. Together with Mount Wilson, the 139 northern Areas stars counts were made up to 2.5 magnitudes (a factor of 10) fainter. In the same year the catalogue was published together with Frederick Hanley Seares (1873–1964). The *Harvard Groningen Durchmusterung of Selected Areas* included some 250,000 stars in all 206 areas; the *Mount Wilson Catalogue of Photographic Magnitudes in Selected Areas 1 to 139* almost 68,000.

In addition to the apparent magnitude of a star, an indication of what kind of star was involved was also needed. This was done via the spectral type, which classifies the absorption lines in the spectrum. Two German observatories were involved there, the Bergedorfer Sternwarte near Hamburg for the north and the Postdam Astrophysikalisches Observatorium near Berlin for the south. This involved the use of a so-called objective prism, a thin prism in front of the telescope's primary lens, which produced for each star a small spectrum instead of a point, enough to determine the spectral type. The Kapteyn Laboratory cooperated with the Bergedorfer Sternwarte. The results were published between 1935 and 1953 for the northern fields as the *Bergedorfer Spektral Durchmusterung der 115 Nördlichen Kapteynschen Eichfelder* and between 1929 and 1938 in the south as the *Spektral Durchmusterung der Kapteyn Eichfelder des Südhimmels*, but of course much material was available earlier to Oort. The third director of the Kapteyn Laboratory, Adriaan Blaauw (1916–2010), was always annoyed by the fact that the northern catalogue was named the *Bergedorfer Durchmustering*, while a large fraction of the work had actually been done in Groningen. With these catalogs, the spectral types of stars up to magnitude 12 (250 times weaker than the naked eye) all over the sky were known. The *Durchmusterungen* together comprise about 231,000 stars.

Distances (parallaxes) were of course of the greatest importance and especially proper motions. Before that much work was done at the Radcliffe Observatory of Oxford University. In the northern areas plates were taken since 1909 under director Arthur Alcock Rambaut (1859–1923), but parallaxes turned out to be not measurable. Each area had been photographed several times, while one of the plates was stored without being developed. Under his successor Harold Knox-Shaw (1885–1970) these were exposed again from 1924 onward. Eventually, the proper motions of some 32,000 stars were measured in *The Radcliffe catalogue of proper motions in the Selected Areas 1 to 115*. A further source for Oort was the work of Peter van de Kamp and colleagues at the Leander McCormick Observatory near Charlottesville, Virginia, where the proper motions of 18,000 stars were measured.

It is clear that this was a lot of material that would take a long time to process, but Oort started to use it for an analysis as Kapteyn must have imagined. Once again, calculators Gerrit Pels and Johannes Kriest did most of the computational labor. Of course, Oort had to take absorption into account. First he therefore excluded areas less than 10° from the Milky Way plane. Furthermore, he estimated the absorption at higher latitudes from counts of extragalactic nebulae, which Hubble had produced over large areas of sky and which he had made available to Oort before publication. The point, of course, then is how to convert galaxy counts into an amount of absorption. For this purpose, Oort used measurements of colors of faint stars in the *Selected Areas*. This enabled him to convert a decrease in the number of Hubble's galaxies into an average change of color of stars due to absorption. But that was not the amount of absorption itself. For this he made use of the work of a young astronomer, Jesse Leonard Greenstein (1909–2002), who after producing a PhD thesis at Harvard now worked at Yerkes Observatory. Greenstein had used photocells to accurately determine the energy distribution in the light of stars as a function of wavelength and had found on that basis that the amount of absorption is inversely proportional to the wavelength. He also compared stars that were just visible through dark clouds of dust with stars next to them that had undergone much less absorption. In the end, Greenstein was able to convert the amount of reddening of a star into an amount of absorption that had caused it. Oort further deduced that the absorption took place in a thin layer around the Milky Way plane, so effectively all stars in a certain *Selected Area* had undergone the full amount of absorption that Hubble's counts indicated.

The next step was to calculate the density along the line of sight for each of the *Selected Areas* (except for those close to or in the Milky Way) from all these data. Of course, that was a huge amount of effort in the form of calculations. Oort found that the density of stars definitely increased in the direction towards the center of the Milky Way and was reasonably symmetrical relative to the plane. Eventually he came up with Fig. 6.17 as the final result of all these calculations, which represents a cross section of the Galaxy. The Sun is in the middle, the center to the right outside the figure. The lines are contours of equal star density. It indicated that the Sun lies between two spiral arms. At higher distances from the plane, the spiral structure disappears and there is only a general increase in density towards the centre (which Oort determined to be by a factor of 1.39 per kiloparsec). We now know that the bright and young stars of type *O* and *B* are concentrated in the spiral arms. We still think the Sun is just on the outside of a spiral arm.

This work was a sort of conclusion of Oort's studies of the structure of the Galaxy using star counts and the program of Kapteyn's Plan of *Selected Areas*. Oorts analysis, which was a culmination of the statistical astronomy that Kapteyn had started, also indicated that this was about the maximum of what could be found out about the structure of our Galaxy at that time.

6.7   The McDonald Inauguration

In 1939 Oort was invited to the inauguration of the McDonald Observatory in Texas. This new observatory was financed by will of William Johnson McDonald (1844–1926), a Texan banker who died childless. He had left most

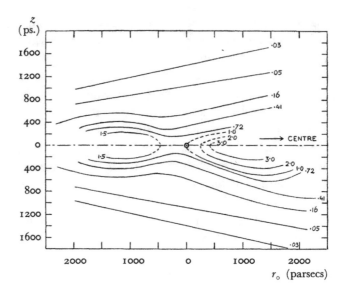

Fig. 6.17 Lines of equal density in the distribution of stars in a cross section though our Galaxy. The Sun is in the center and the Galactic center to the right outside the figure, according to insight of that time at 10,000 pc

of his fortune to the University of Texas to build a large telescope in Texas. This was unexpected and was contested with some success by several family members, but eventually the university received 800,000$ (in current purchasing power about 12.5 million $), enough to build a large telescope. This was in 1933. One problem was that the university did not even have an astronomy department! But this was resolved by setting up a cooperation with the University of Chicago and Yerkes Observatory. According to the agreement it was determined that for thirty years McDonald would bring in the telescope and Yerkes the astronomers. When the agreement expired in 1963, the observatory became part of the by then thriving astronomy department of the University of Texas at Austin.

The location was determined on Mount Locke, a 2,070 m high mountain in southwest Texas near the border with Mexico and close to the town of Fort Davis. This was about 800 km from Austin. The first director was Otto Struve (1897–1963), descendant of the von Struve family of astronomers, directors of the Pulkovo Observatory near Saint Petersburg. Struve had dropped the 'von' during his naturalization as an American citizen and was now director of Yerkes Observatory. He supplied the specifications for the telescope, which got an 82 in. (2.1 m) primary mirror. It became the second largest telescope in the world and, also thanks to the dry climate of Texas, a very productive instrument.

Fig. 6.18 The Oort family in the garden behind their home at the Sterrewacht. This photo dates from 1939 or 1940. From the collection of Abraham H. Oort

The inauguration on 5 May 1939 was celebrated with a symposium to which the most prominent astronomers were invited. Struve had invited Oort to spend two months at Yerkes Observatory in combination with this. That would have to be afterwards because Hertzsprung would be in South-Africa for some time and Oort could not get away from Leiden before the ceremony. But Oort declined this offer and seized the opportunity to carry out another plan, namely an observing session at Mount Wilson to collect more photographic plates of extragalactic systems. The few plates Oosterhoff had taken in 1935 had proved to be very useful and Oort wanted to continue his program, which he had attempted on Perkins Observatory. Going to Yerkes after that would be too late for his lectures in Leiden. Oort left on April 18, 1939; his trip would last four and a half months and Mrs. Oort had to stay at home with children of now ten, eight and four years old (see Fig. 6.18). The outward journey was on a German ship, boarding in Cherbourg. It was very quiet because few dared to travel on German ships, but the crew was extremely courteous. No attention was paid to it, when it was Hitler's birthday during the crossing.

After a visit—as always—in New York to the Schilts, Oort traveled by train to Alpine, Texas, the railway station closest to the new McDonald Observatory. There he was met by Gerard Peter (actually Gerrit Pieter) Kuiper (1905–1973) and his wife. Kuiper had studied in Leiden and had obtained a PhD under

Hertzsprung in 1933. with a thesis on double stars. He first worked at Harvard, where he married Sarah Parker Fuller (1913–??) in 1936 and was now attached to Yerkes. Kuiper had done most of the organization of the ceremony and symposium.

Oorts presentation during the congress would be a new milestone. In the first part he dealt with the 'deviation of the vertex'. Fundamental dynamics dictate that in the symmetrical Galaxy the random velocities in the Sun's environment must be largest in the direction towards and away from the center. However, there was a well-established deviation of 20°, already known from the fact, that the direction of the Star Streams of Kapteyn were this amount different from the later determined direction to the Galactic center by Oort from the rotation. Oort showed with some simple numerical simulations that this deviation could easily be explained by irregularities in the gravitational field caused by spiral arms. The second part of his contribution was the most impressive.

The photographic plates that Oosterhoff had taken for Oort at Mount Wilson had yielded good results. Herman Kleibrink had done most of the work again. The results as presented by Oort are shown in Fig. 6.19. The unit along the (logarithmic) vertical axis ($\log I = 0$) is 25 magnitudes per square second of arc. That means that from a square of 1 by 1 arcsecond as much light is seen as from a star with magnitude 25. Now a star of magnitude 6 is just visible with the naked eye and 2.5 magnitudes correspond to a factor of 10. So this would be a star that is some 19 magnitudes, or 40 million times fainter than we can see. A dark sky like at the best observatories has a surface brightness of about 22.5 mag/arcsec^2. The faintest measurements in Fig. 6.19 are about this unit of 25 mag/arcsec^2, so that is roughly one tenth of the brightness of the background sky.

Before discussing this further it is good to say a little more about the technique of photographic surface photometry. Although in principle straightforward, in practice it is not so simple. Part of the problems comes from the photographic emulsion itself. The dynamic range is limited because it may be underexposed or overexposed. Furthermore, the emulsion must be very uniform in sensitivity, especially if faint outer parts are measured. The inner parts of a galaxy are bright, but further out the surface brightness of the disk is fainter than the sky itself. So the exposure must be such that the sky background is not under-exposed. That means that a large telescope and long exposures are required. Then the trick is to measure the small increase in total surface brightness compared to that of the sky itself, as is registered immediately next to the galaxy. Oort was able to measure brightnesses in a system at a level to 10% or so of the sky. That was definitely state of the art. But it remained a very tedious and time-consuming business. It has only become possible in the early 1980s

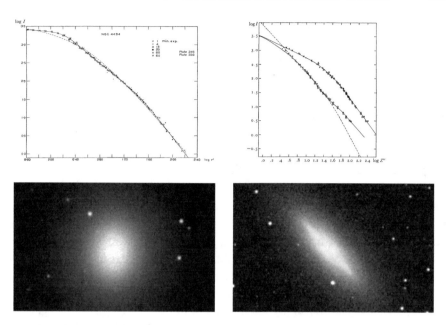

Fig. 6.19 Top: The light profiles (surface brightness as a function of distance from the center) of two galaxies. The different symbols correspond to different quadrants. On the left NGC 4494, an elliptical system with moderate flattening (for which the profile along the minor axis has been corrected). On the right NGC 3115, an S0 system with a disk but no spiral structure. The lower measurements are along the minor axis, the upper measurements along the major axis. The dotted line is the de-projected brightness distribution along the major axis. Diagrams at the top from the Oort Archives. The pictures of these systems at the bottom have been produced with the Digital Sky Survey [76]

to do on a routine basis. A first large-scale survey was done by myself and my collaborators with a measuring device of the Sterrewacht Leiden (called Astroscan) that could scan plates under computer control. This was then followed by processing in Groningen with a specially developed software package. The faintest measurable surface brightnesses (using photographic plates obtained with the 48 in. Schmidt telescope at Palomar Mountain, California) were of the order of 1 percent of that of the sky. Nowadays electronic detectors (CCD, charge-coupled devices, as in digital cameras and telephones) are used and it is possible to measure much fainter surface brightnesses. However, Oort's results, as presented at MacDonald in 1939, were astonishingly good for that time.

Oort's analysis of the data on NGC 3115 was the most detailed. This is an S0 galaxy (S-zero), a transition type between an elliptical and a spiral galaxy, because it does have a disk like our Milky Way Galaxy, but without spiral structure, and just like elliptical galaxies, no dust or bright, young stars (and no gas as we know now). That this disk is seen from its side can be appreciated from Fig. 6.19. Oort showed that if one assumes that the system is symmetrical, one can deduce from the two-dimensional measurements what the three-dimensional distribution of stars is. That requires quite some calculation, again done by Gerrit Pels. Oort assumed that this was also the distribution of *all* matter, so that the gravitational field could be calculated. Except for one factor, which we nowadays call mass-to-light ratio (I will use the notation M/L). This factor connects in a certain small volume the amount of starlight (expressed in the luminosity of the Sun) L to the mass M (in the mass of the Sun). Then one can use the equations of galactic dynamics to find out what the velocities of the stars should be in order to maintain equilibrium.

Measuring velocities is much more difficult. After all, you have to disperse the light in wavelengths and then also register the absorption lines in the spectra of the stars. And we already saw that the total light, except in the very central parts of a galaxy, is fainter than that of the dark night sky. I will illustrate this with the story of how the rotation of galaxies was discovered. This happened in 1914 independently by Vesto Slipher at the Lowell Observatory in Flagstaff, Arizona, and by Maximilian Franz Joseph Cornelius Wolf (1863–1932) of the Landessternwarte Heidelberg-Konigstühl. Slipher observed the very bright central part of the Sombrero nebula (Fig. 6.5). His exposure lasted 80! hours, of course spread over several nights. In October 2014 I attended a conference at Lowell Observatory, where some of us were taken on a visit the vaults with photographic plates. Besides the plate on which Pluto was discovered in 1930, we also saw the spectrum that Slipher took. My colleague Bruce Gordon Elmegreen (b. 1950) photographed it through a small magnifying glass (Fig. 6.20). Even with the contrast enhancement, which I have applied, the rotation is barely visible. Reasonably reliable measurements of motions of the stars in the central parts of elliptical systems only date back to the 1960s and 1970s, especially when electronic detectors became available. In the much fainter disks of galaxies stellar motions were measured only in 1983 when I, together with my Australian colleague Kenneth Charles Freeman (b. 1940), made exposures of an entire night on one of the largest telescopes (the 3.9 m Anglo-Australian Telescope of Siding Spring Observatory). We used an ingenious detector developed by Alexander (Alec) Boksenberg (b. 1936)—later director of the Royal Greenwich Observatory—to record the arrival of individual photons. It is a weird experience to see individual photons arrive on a electronic screen.

In 1939, Oort had some measurements at his disposal that Milton Lasell Humason (1891–1972) had obtained at Mount Wilson Observatory. He had measured a rotation velocity at four points along the major axis, which increased to 450 km/s at 45 arcseconds from the center (about halfway Fig. 6.19). In the center he had estimated from the width of the lines that the velocity dispersion

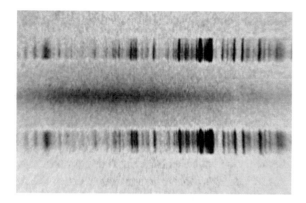

Fig. 6.20 Spectrum on which Slipher discovered the rotation of the Sombrero Nebula (Fig. 6.5). Along the horizontal direction wavelength and vertically the position in the sky along a narrow slit parallel to the dust layer. Above and below the galaxy spectrum the spectral lines from a gas discharge lamp (usually with argon and/or neon) for wavelength calibration. Lowell Observatory Archives (with permission); photo taken by Bruce Elmegreen (see text)

(a measure for the average random velocities) could be a few hundred km/s at most.

From the measured rotation and light distribution Oort estimated that the M/L had to be of the order of 150 and probably even more. That was strange, since for the neighborhood of the Sun this was only about 2. Next he considered the vertical distribution of light and assuming that mass-to-light ratio would be 200, he found that the vertical velocity dispersion at two of the distances where Humason had measured the rotation should be 200 to 250 km/s. Then he turned to the distribution in the plane. First, he calculated the rotation velocity in a circular orbit using the same M/L. That, however, is not directly comparable to what was measured for two reasons. Firstly, the measurement is an average along the line of sight through the galaxy and the rotation has an component along the line of sight that varies along it. Secondly, by now it had become clear that there was such a thing as 'asymmetric drift'. This can be understood as follows. When stars move in pure circular orbits, their orbital velocity corresponds to a centrifugal force that compensates the gravitational force exactly. This is called the circular velocity. But in practice, stars on top of rotation have random motions. The effect of asymmetric drift is that a group of stars then collectively lags behind that circular velocity. You can also understand that by realizing that there is kinetic energy in the random motions and therefore less energy is needed in the systematic motions to compensate the potential energy in gravity. So the larger the random motions (the velocity dispersion), the greater this lag will be. This was found by Gustav Strömberg (who was therefore not entirely wrong when he contradicted Oort that the asymmetry in the high-velocity stars could also be seen in low velocity stars), and by now Oort had theoretically explained it in his study of the dynamics of the solar neighborhood.

In his analysis for NGC 3115, Oort corrected for these effects, taking the horizontal velocity dispersion to be the same as he had estimated in the vertical direction. After all, dynamic studies show that in symmetrical systems the dispersions in the vertical and radial (towards the center) directions must be equal. Only when the system is strongly flattened (like our Milky Way Galaxy) close to the plane, the two motions become independent. Oort could not in any way reconcile the observations with the calculations. We now know this is because Humason's rotation velocities were seriously in error. Instead of rising to 450 km/s, the rotation velocity quickly flattens out at about 200 km/s and remains at that level up to reasonably far out. Measuring the stellar velocities in galaxies was not possible at that time. It was just too early to do this kind of work. But Oort's analysis did indicate how to go about it, and that was a very important step.

6.8   Mount Wilson

In his letters to his wife, Oort mentioned that Struve had offered him the position of director of Yerkes Observatory. Struve himself would then remain director of the astronomy department in Chicago and of McDonald Observatory. That was a very attractive offer, because it would give Oort direct access to the telescopes of both Yerkes and McDonald. And he also alluded to the fact that leaving Europe now with the threat of war with Germany was also an attractive idea. However, Oort immediately indicated that he probably would not accept this; a factor that played a role was that he did not want to abandon Leiden in these times, in addition to Mrs. Oort's reluctance to emigrate.

After the conference at McDonald Observatory, Oort traveled to Pasadena. There were several people who also made that trip, including Bertil Lindblad. But also some staff members of Mount Wilson, especially Edwin Hubble and Walter Baade. With the latter Oort over the years would have much contact and collaborate. Wilhelm Heinrich Walter Baade (1893–1960) was a German astronomer, who worked on Mount Wilson on nearby galaxies. They talked for a long time during the train journey. Baade would help Oort on Mount Wilson with the observing that was planned.

Oort stayed in California from May 10 to June 25, where he had several observing runs on Mount Wilson with the 100 in. and 60 in. telescopes. The weather did not really cooperate, but fortunately Baade gave him three more of his own nights on the 60 in. The annual report of the Sterrewacht mentioned

Fig. 6.21 Oort with three astronomers that were originally Dutch, but have worked in the United States most of their lives. The picture has been taken in 1939 at Harvard. From left to right: Bart Bok, Oort, Peter van de Kamp and Jan Schilt. From the Oort Archives

that Oort took 122 plates. This would form the basis of an extensive study of light distribution in extragalactic nebulae, although the final result would have to wait until 1961.

The return journey home was again in several stages. First Oort visited the Lick Observatory near San Francisco (and made a trip to Lake Tahoe, where an earlier acquaintance lived, an American professional in the Army, who had written a PhD thesis in physics in Leiden) and then to Yerkes, where he spoke to Struve again and definitively rejected the offer of a directorship. He spent some time at Harvard (Fig. 6.21). At Yerkes Oort gave presentations and lectures, undoubtedly partly to finance the trip. At Harvard Bartholomeus Jan (Bart) Bok (1906–1983) worked, who had studied in Leiden and obtained his doctorate in Groningen. He married Priscilla Fairfield (1896–1975), who was also an astronomer. Bok later became deputy director of Harvard Observatory, director of the Australian Mount Stromlo Observatory near Canberra in 1957 and of Steward Observatory in Tucson, Arizona from 1966.

But then it was time to go back to the Netherlands. After a visit to Jan and Jo Schilt in New York, Oort traveled back, not with a German ship this time, but with the SS Statendam of the HAL. On September 2, he was home again. The day before, the Second World War had broken out with the invasion of Poland.

Fig. 7.0 Oort in 1939. From the Oort Archives

7

The Watershed: World War II

War is not one disaster for millions.
War is millions of times a disaster for one.
Henricus Antonius Franciscus Maria Oliva (Hans) van Mierlo (1931–2010).

The Second World War made a deep cleavage,
not only in the <u>lives</u> of our generation, but also in the development of astronomy.
Looking back it appears as a wall separating two different astronomies.
The great change was mainly brought about by the advent of radio astronomy.
Jan Hendrik Oort.

7.1　First Years of the War

In the Netherlands, the Second World War began in the night of May 9 to 10, 1940 with the invasion of the German army. Oort was in Utrecht that night according to his pocket diary. From the 10th of May he wrote down his experiences there with ink; here and there you could still see notes with pencil in the background. For May 10 it reads in pencil: 'Second Regiment Field Artillery' and 'Draft 1923'. According to his military passport, Oort was transferred there in 1938. The regiment's barracks were in Ede, not far from Utrecht, so probably Oort would have reported there. He had traveled by bus to Utrecht in the afternoon of May 9. But apparently he did not have to stay,

---

Hans van Mierlo was a Dutch politician.
From *Reminiscences of Astronomy in the Twentieth Century*, Oort (1982).

© The Editor(s) (if applicable) and The Author(s), under exclusive license
to Springer Nature Switzerland AG 2021
P. C. van der Kruit, *Master of Galactic Astronomy: A Biography of Jan Hendrik Oort*,
Springer Biographies, https://doi.org/10.1007/978-3-030-55548-1_7

Fig. 7.1 The Oort-family in 1940 in Leiden in the back of their house at the Sterrewacht. From the Oort Archives

because on May 10 he returned to Leiden. It was a scary time. At the end of the week, in the 'Memorandum' box, Oort wrote: 'Bramke is packing a package for the 'prison'.' Bram Oort was then five years old and terrified (Fig. 7.1).

In the beginning everything at the university went on normally (for a historiography of Leiden University during the war, see [78]). The first sign of trouble came on October 21, 1940 with the requirement that all professors and other personnel had to fill in an 'Ariër verklaring' (Aryan statement). In this one had to indicate whether parents and grandparents (of the person his/herself and his/her spouse) were of Jewish descent. A meeting of the Senate (the collection of all professors), to discuss how to deal with it, was forbidden. The plan to have this meeting had probably been passed on to the occupational forces by Nazi collaborator Jacobus Johannes Schrieke (1884–1976), extraordinary professor in the Faculty of Law and later Secretary-General of the Ministry of War during the occupation. Because meetings of more than twenty people were forbidden, a small group formed from the Senate, which secretly gathered to discuss matters like this one and others. This was called the 'Kleine Krans' (Small Circle). There was no common response formulated to the question of the Aryan Statement; people signed (or would lose their jobs), often accompanied by a note protesting against the state of affairs.

The Oort Archives contain extensive notes about the times during the war in diaries and pocket diaries, as well as various notes and newspaper clippings about this period. Including a piece written after 1945 entitled 'Stories of the Netherlands in wartime'. Documentation available is not conclusive about the question whether or not Oort was a member of this 'Kleine Krans', but there seems little doubt about it. There was a meeting on February 8, 1941 with a dinner on the occasion of the dies natalis (anniversary) of the university, of which the menu is still available (I am grateful to Professor Otterspeer for making this available to me [79]), at the home of Berend George Escher (1885–1967), professor of geology. The printed menu included a woodcut of his famous brother, Maurits Cornelis Escher (1898–1972), of a fence closed with a padlock and the text: 'An image of bad faith, obstructing all it can; once bred in discord, raised in suspicion'.[1] Among the fifteen guests were professors Jan Julius Lodewijk Duyvendak (1889–1954), oriental languages, Roeland Duco Kollewijn (1892–1972), law, historian Johan Huizinga (1872–1945) and physicist Hendrik Brugt Gerhard Casimir (1909–2000). This could be a part of the Kleine Krans. Rector Magnificus Alexander Willem Byvanck (1884–1970) was not present. Of course such meetings were very secret, and that is probably why there is no indication of this in Oorts pocket diary; only 'storm and strong thaw'. Two days before that Oort had skated a large part of the 'Eleven-city tour' (this about 200 km tour calls at each of the the eleven Frisian cities). In the written piece quoted above 'Stories' Oort refers to a meeting of the Kleine Krans on April 28, 1942 (coincidentally Oort's birthday), but the pocket diary with appointments only mentions 'bios [short for bioscoop, movie theater] at Boschma'; apparently encoded notes were used.

After the Aryan declaration, things became rapidly more serious for Jews. On November 21, 1940, all Jewish professors were dismissed. One of them was Eduard Maurits Meijers (1880–1954), professor of Dutch and international law. His former student Rudolph Pabus Cleveringa (1894–1980) was at that time dean of the Faculty of Law and he gave a lecture at the time of Meijers' next scheduled lecture, Tuesday November 26, 1940. In it he protested strongly against the dismissal of Meijers and other Jewish professors. This took place in

---

[1] 'Van kwade trouw het beeld, versperrend naar vermogen; in onmin eens geteeld, in achterdocht getogen.'

the auditorium of the Academy Building and according to his pocket diary Oort was present: '10 o'clock, lecture Cleveringa'. After the lecture was concluded, students spontaneously started singing the 'Wilhelmus', the national anthem of the Netherlands. The next day Oort wrote 'university closed' in his pocket diary. Some students distributed stencils of the text of Cleveringa's lecture [80]. Oort kept a copy in his Archives. To date, Cleveringa lectures are organized annually all over the world by the University of Leiden to commemorate this courageous act. Cleveringa was arrested by the Germans but released in the summer of 1941, and taken hostage in 1944. He survived the war. Meijers was arrested in August 1942 and survived several concentration camps.

From the 'Stories of the Netherlands in wartime':

> '25 Nov. '40. Meijers receives letter of resignation. Cleveringa: 'I will not attempt with my words to lead your thoughts to those that have produced this letter. Their act qualifies itself. The only thing I desire now is to keep him out of sight for those below us and to turn your gaze to the height where the glowing brightness of him stands, what we are here for'. Possessing the transcript is reason to be punished with a month's imprisonment. [...]
>
> In Aug. '41 it was announced that some Leiden professors will be fired and replaced by NSB sympathizers. Kleine Krans recommended in this case to resign en bloc. It was around October 3rd. [...] This plan made President-Curator abandon his plans for the time being.'

The NSB is the Nationaal-Socialistische Beweging (Movement), the party that supported the Nazi's.

> 'April '42 [...] plans in Leiden were implemented anyway. [...] The problem is always that the steps were so small (also in Germany at the time). So always hesitant whether this was the right time. [...]
>
> Kleine Krans at Kollewijn 28th April '42. In any case, those present committed themselves to resign and to involve as many others as possible. At the beginning of May, 80% resigned. Rather easily accepted by the Germans. About twenty professors were singled out and taken to hostage camp. Released later, however, but with a ban on returning to Leiden.'

From 1942 onward, the Germans took leading civilians such as professors, politicians, preachers, artists, writers, etc. hostage and took them to the internment camp Sint-Michielsgestel near Eindhoven. They might be killed in retaliation for sabotage and other acts of the resistance movement. The first group of 460 people arrived in May 1942, a maximum of 700 people were interned at the same time. Kollewijn had been taken hostage in 1941, but was banned to Hoogeveen because of his health. The meeting must have taken place without him.

The Utrecht astronomer Marcel Minnaert was interned in Sint-Michielsgestel from May 1942 to April 1944. He taught physics to his fellow prisoners. There was good reason for Oort to fear the same fate and in the summer of 1942 he and his family planned to go into hiding elsewhere in the country.

7.2   Personal Matters and Scientific Research up to 1942

Many things of private nature went pretty normal in the beginning. The winters were harsh and there was a lot of skating. The Eleven-city tour was held for three consecutive years in the very cold winters of 1940, 1941 and 1942. On Monday January 29, 1940 (so still before the outbreak of the war) Oort traveled to Friesland over a snow-covered *Afsluitdijk* (Closure Dam) and skated the next day in the Eleven-city tour. He skated together with a good friend, Anton Bicker Caarten (1902–1990), who was an expert in Dutch windmills and had written notable books about them. From Oorts pocket diary:

> 'Tuesday, January 30th. Eleven-city tour. Sneek, IJlst, Sloten, Stavoren, Hindelopen, Workum, Bolsward, Harlingen with Anton Bicker Caarten. Back by train because there was too much snow on the Afsluitdijk.'

So nine of the eleven cities (city of the start Leeuwarden included) and 116 of the 199 km. This was probably the first time Oort took part in the tour. The earlier ones had been in 1927 and 1933. For precisely those years his pocket diaries are missing from the Archives and there is no reference to them in his diaries. Oort's skates, at least probably the most recent ones he used, were 'Friese doorlopers' (see Fig. 7.2). The following year Oort skated the tour again with Bicker Caarten. They traveled by train; in the pocket diaries there is a schedule via Utrecht and Zwolle to Leeuwarden, departure from Leiden on February 4 at 14:14 and arrival at 20:02. The next day they attended a concert.

> 'Thursday, February 6: With Anton Bickers Caarten. Eleven-city tour. Left Leeuwarden at 6:30. It was still dark and 14 [degrees] below zero. Until Dokkum and back to Barthlehiem in the morning twilight; from there via Vrouwbuurtstermolen and Berlikum to Franeker, Harlingen to Sloten, where I still have 68 km to go at 6 o'clock. For safety's sake, I choose to go back by train. Anton continues with two farmer's sons called Schipma and arrives in Leeuwarden at 11 o'clock.'

This time the direction of the trip was anticlockwise, in contrast to the previous year. For 1942 Oort does give a list of skating tours (among other

Fig. 7.2 Oort's skates, donated by the family to Leiden University. They are now part of the collection of the Academic Historical Museum in Leiden. They are 'Friese doorlopers', as the word suggests originally from Friesland. The word 'doorloper' (from the verb 'continue') refers to the property that the 'iron' (the metal part made of a hard kind of steel) continues all the way to the end of the wooden frame under the heel. From the Oort Archives

things that Coen skates an 80 km tour with his new 'doorlopers'), but Oort himself does not participate in the Eleven-city tour.

'Today the Eleven-city tour takes place. Here it freezes 15°, in Friesland a few degrees more. 5200 participants. Anton and Jelle are going without me. I have given up my plan at the last minute because I do not feel well enough to make it and my weight has dropped so low that I will get supplementary food. (56 KG). They both make it, but Jelle is exhausted and towed along the last 35 km.'

Jelle was Oort's brother-in-law Jelle Nauta, his sister Emilie's husband; according to family stories Jelle burned his skates after this and never skated again. It has disappointed Oort that he never completed the Eleven-city tour;

but he must have been proud that his youngest son Bram did complete it in 1956 (finished at 19:45!).

Summer holidays were still possible, in addition to the weekends and so in the family's second home in Katwijk. In 1940 the Oorts spent a longer period of time in a certain place called 'Westerflier, Hoendelo'. On old staff maps there is indeed a note Westerflier 1 km west of Hoenderloo (with double o), but there is no building there now. Probably there used to be a holiday home at this spot.

On May 12, 1941 Oort's father Abraham Oort died at the age of 72. The advertisement in the Leidsch Dagblad (the local newspaper in Leiden) only gives the date of the funeral. The Oort Archives contain some documents on the formal settlement of finances and so. Oort senior had retired in 1934. He was an unassuming man. I have not been able to find any obituary or other article about him, except the short note on his 40th jubilee [23].

After the closure of the university, education officially came to a standstill. It is true that there were students who worked at the Sterrewacht, but there were no exams and certainly diploma's awarded. Communication with foreign countries was also limited and certainly the exchange of observational data was impossible. American publications did not reach the Netherlands or did reach the Netherlands only with much delay. Oort managed to stay in contact with Bertil Lindblad in neutral Sweden. In the United States, Bart Bok had been put in charge of the 'Committee for the Distribution of Astronomical Literature' (CDAL), established in 1940 by the American Astronomical Society (AAS), to maintain the exchange of astronomical literature between the United States and occupied Europe. For this they used intermediaries such as William Otto Brunner (1878–1958) in Zürich, director of the Eidgenössische Sternwarte, Bertil Lindblad, and others [81]. Through this chanel Oort must have heard of the death of Frank Schlesinger in July 1943. Oort wrote an in memoriam in 1944, which was not published until 1945 in *Hemel & Dampkring* [82]:

> ... there was a lack of sensational developments, there was a lack of discovery. Schlesinger deliberately refrained from it. He felt it was his vocation to bring forward classical, positional astronomy. [...] Smiling, he once said to me, having entertained the plan when he came to Yale Observatory to have painted above the door: 'He who enters here forsakes all hope of discovery'.'

Oort's scientific work was limited to observatonal material that was available in catalogs or as previously obtained information. Using data that had been published before the outbreak of the war, Oort worked on a better determination of the rotation velocity of the solar neighborhood around the Milky Way center. This was previously determined as the average velocity relative to the globular clusters, but radial velocities were also measured of RR Lyrae stars that occur in the space between the globular clusters. These are late stages of stars that are unstable and pulsate giving rise to a period of light variation, which is a measure of their luminosity so that the distance can be determined. Together with student Adrianus Jan Jasper van Woerkom (1915–1991) a rotation velocity of 240 km/s was found, which is very reasonable compared to what we know now.

Oort published in the *Bulletin of the Astronomical Institutes in the Netherlands*, which was published regularly until 1942. With another student, Johannes Jacobus Maria van Tulder (1917–1992) a better determination of the pole, and thus the plane, of the Milky Way was carried out. Van Tulder left astronomy later. There were some more studies but the lack of access to new observation material limited their depth.

7.3   Nova Persei 1901 and the Crab Nebula

Before the war, Oort had started work in a completely different subject, namely exploding stars. The first piece of work concerned a so-called nova, a new star, but in fact a faint star that had become brighter. This occurs in late stages of stellar evolution and is short-lasting (but the phenomenon may recur). Nova Persei (in the constellation Perseus) appeared on February 2, 1901. While at first only a very weak star had been visible, now all of a sudden it became a star of a brightness comparable to the brightest stars. Max Wolf of the Heidelberg Landessternwarte in August found nebulosities around the star. This was confirmed by Charles Dillon Perrine (1867–1951) of Lick Observatory and George Ritchey (1864–1945) at Mount Wilson. These nebulosities were regularly photographed from September 1901 onward (see the upper panels of Fig. 7.3, which are drawings based on Ritchey's photographs). Soon it was found that these nebulosities showed expansion, which would be incredibly fast unless the star was very close to us. Kapteyn had given an explanation for this, namely that it was not material that we see expanding, but the light-front of the relatively short peak in brightness during the outburst, which now moved along reflecting structures around the star. That would then happen at the speed of light of 300,000 km/s and on that basis Kapteyn estimated that the star had to be at a distance of 300 lightyears (or 90 pc). The idea was quickly confirmed by observations by Charles Perrine, which showed that the nebulae and the star had the same spectrum and that it therefore had to be reflected light.

Nova Persei 1901, renamed GK Persei nowadays, continued to be observed. In 1916, Edward Emerson Barnard (1857–1923) found that a weak nebula

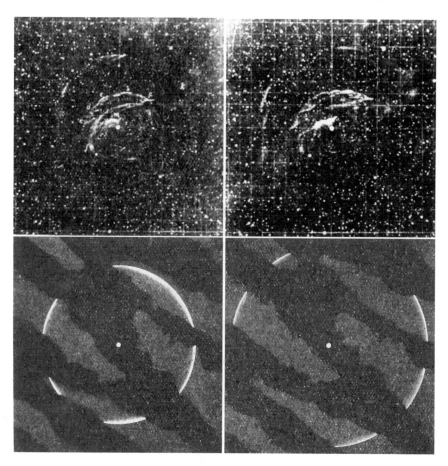

Fig. 7.3 Top: drawings from photographs by Ritchey of Nova Persei 1901, on the left September 20, on the right November 13, 1901. Bottom: the explanation of Kapteyn in a schematic drawing of the expansion of the light front (on October 1 and on December 1). From the Oort Archives and [83].

had formed around the star, which now turned out to be the star's expanding outer parts. If Kapteyn's distance were correct, it would indicate an expansion rate of 400 km/s. But in 1935 Milton Humason had recorded a spectrum, which showed that that velocity was actually of the order of 1200 km/s. This was remarkable, because it would indicate that the distance had to be larger than Kapteyn had inferred and the apparent expansion velocity of the filaments three times the speed of light.

Oort came up with an explanation for this, namely that the light front did not run along reflecting matter around the nebula, but was projected on a thin structure at a greater distance. Suppose there is a screen between the star and us perpendicular to the line of sight. At a certain point, the expanding front of

Fig. 7.4 The Crab Nebula, photographed with the Hubble Space Telescope [84]

light, which is a spherical shell because the bright period has been short, will intersect this along a circle. So we see a circle lighting up on that screen. But at a later moment that circle has grown and if the distance from the screen to the star is large enough, it has expanded faster than with the speed of light. Matter cannot go faster than light, but this circle can. Or as Oosterhoff explained it in the elementary astronomy lecture course that I attended as a student: think of a lighthouse with a large screen around it. Then the spot of light moves across that screen and nothing can stop us from putting that screen so far away that that spot goes faster than the speed of light.

Before Oort would publish this, he discovered that Paul François Jean Couderc (1899–1981) had independently published this explanation in France. Oort refrained from further publication and left the honor to Couderc. He did give lectures about it during the war in the Netherlands for astronomers and in 1943 Jean Jacques Raimond (1903–1961) wrote it down for an article (in two parts) in *Hemel & Dampkring* [83][2]. Raimond had studied in Groningen with van Rhijn and had obtained his PhD in 1934 for work on interstellar absorption. For many years he was director of the Zeiss Planetarium in Den Haag.

---

[2]See Appendix B4 of [2] for an English translation.

Fig. 7.5 Jan Julius Lodewijk Duyvendak, professor of Chinese language and literature at Leiden University. From Wikicom-mons [85]

In the early years of the war, Oort also completed work on the identification of the Crab Nebula (see Fig. 7.4). The Swedish astronomer Knut Emil Lundmark (1889–1958), who worked at Mount Wilson Observatory for several years, had found in ancient Chinese chronicles that in 1054 a guest star had been observed at about this position, which was so bright that it remained visible during the day. John Charles Duncan (1882–1967) had measured that the image of this nebula in the sky was expanding and that it must be of the order of 900 years old. Slipher had already in 1916 found lines in spectra that also indicated expansion, and around 1940 Nicholas Ulrich Mayall (1906–1993) at Lick Observatory had repeated that and had measured the velocities more accurately. On that basis a distance of about 1.5 kpc was deduced (nowadays we think it is about 2 kpc). Oort was immediately interested in the brightness of the star during the explosion. We now know it was a supernova. This is an exploding star, sometimes a single star at the end of its existence, sometimes in a double star when there is mass transfer between the two components. Jan Duyvendak (Fig. 7.5), whom we already met at the Kleine Krans, made an extensive study of the Chinese chronicles at Oort's instigation. On that basis, more was learned about how bright the star had been in 1054 and how it had changed in the months that followed. On this basis, it was concluded that the absolute magnitude during the maximum would have been -16.5, i.e. something like 250 million times more luminous than the Sun (Fig. 7.6).

Fig. 7.6 The Crab Nebula is the spot near the bottom-right corner close to the bright star ζ Tauri. The text near the arrow reads: 'guest-star'. The text is (without detailed time indications): left column: 'On [1054, 4 July] a guest star appeared several inches [thumbs] south-east of [ζ Tauri]. After more than a year it gradually became invisible.' And in the column on the right: 'On [17 April 1056], the head of the Astronomical Bureau reported that as of [9 June–July 8, 1054] a guest star appeared in the morning in the eastern sky, which remained in [ζ Tauri], and has only now become invisible. Translation by Duyvendak. From *Honderd jaar Leidse Sterrewacht* [45]

These findings were published in the United States in 1942 in the form of two articles, one by Duyvendak and another by Mayall and Oort. Mayall had arranged this, because in 1941 material could still be sent to the United States. Duyvendak and Oort saw these publications themselves only after the war.

7.4 Astronomers Conferences

In 1941 the first Dutch Astronomers Conference was held. The Dutch Astronomers Club, the professional association of Dutch astronomers, had existed since 1918, but it met a few times a year for only half a day or so. The conference, organized by Marcel Minnaert, was aimed at bringing together astronomers and students for a somewhat longer period of time. It lasted from lunchtime on June 3 until about the same time on June 5 at a location in Doorn not far from Utrecht. These conferences still exist, now mostly held in the month of May, and together with Flemish astronomers and students. The group photo of the first conference is shown in Fig. 7.7.

Fig. 7.7 Participants in the first Dutch Astronomers Conference in 1941. Oort is on the far left, next to Gijs van Herk and Marcel Minnaert. In the back to the left of the sundial with beard and pipe Jaap Houtgast. On the other side Adriaan Blaauw and in front of him (looking down) Lukas Plaut. The second person from the right is Kees de Jager. In the front left Fjeda Walraven, and next to him Henk van de Hulst. From the Dutch Astronomers Club (NAC) [86]

Fig. 7.8 Left: Oort during his lecture at the first Dutch Astronomers Conference in 1941. The number 4594 on the blackboard is the catalogue number (NGC) of the Sombrero nebula in Fig. 6.5. On the right: Pieter van Rhijn en Oort during the second Dutch Astronomers Conference in 1942. From the Dutch Astronomers Club (NAC) [86]

The conference was held again in 1941. Minnaert was missing because he was taken hostage to internment camp Sint-Michielsgestel. At both conferences Oort had extensive presentations. The directors gave overviews of the work at their institutes; Oort did this for Leiden on behalf of Herztsprung. As a Dane he kept himself deliberately and carefully in the background so as not to antagonize the Germans. He quietly did his research and interfered as little as possible with everything else. Oort talked about galaxies and international cooperation at the first conference and at the second about fundamental reference systems (for the celestial positions of stars). In addition to presentations, there was the possibility of playing music, taking walks and playing soccer and table tennis. Fig. 7.8 shows Oort at the conferences.

7.5   The Potbrummel at Hulshorst

From May 1942, the Germans had arrested a large number of prominent civilians, including professors, and took them as hostages to the internment camp Sint-Michielsgestel. Most of the members of the Kleine Krans, at least the fifteen who were invited to the dies natalis dinner in February 1941, had resigned from their professorships, most of them around June 1, 1942. Escher, Duyvendak and Kollewijn, for example, were taken to the camp; Huizinga was

Fig. 7.9 The 'Potbrummel' at Hulshorst near Nunspeet. This is the cottage where the Oorts lived during the second half of World War II. Photograph made available by M.J.A. (Marc) Oort.

arrested a little later in August, but he was released again for health reasons. Casimir had escaped this fate because he went to work at the Philips Physics Laboratory (the famous Philips NatLab, which existed and was supported by the Philips company until 2001) in Eindhoven. This institution was seen as strategically important for Germany and staff there was spared. Throughout its existence it has been world renowned for fundamental research in electronics, physics and chemistry, computer science and information technology. Casimir was director of Philips NatLab from 1946 to 1972.

Oort was of course worried about being detained as well and decided to go into hiding outside the western part of the country. He found an opportunity in a vacation home (see Fig. 7.9) owned by Henri Bienfait (1900–1990), a chemist who had studied and obtained his PhD in Amsterdam. How he had come to know him is not clear; Bienfait worked his whole career at Philips NatLab (and eventually became deputy director), so maybe Casimir had played a role in this. Bienfait's wife was Bertha Elisabeth van Osselen (1902–1977) and the cottage was located on the grounds of the home of her parents, Jan Rudolph van Osselen (1869–1948) and Johanna Henriëtte Spakler (1872–1955). Van Osselen had worked in the Navy and had later become director of a mortgage bank until his retirement. The location was in the village of Hulshorst near Nunspeet on the Veluwe. The address was Potbrummel, and the house itself was called Potbrummel also. Brummel is local dialect for blackberry.

Oort first went to see it on Monday, July 20, 1942. His pocket diary contains notes to remind him to inquire about voltage 'for the radio, toaster and vacuum cleaner'. Oort had also resigned, although it had not yet formally taken effect, and there was the question of how to generate income. Through his father-in-law, Willem Graadt van Roggen, Oort had received an invitation to have a conversation with the director of the N.V. Nationale Levensverzekeringsmaatschappij, a large life insurance company, where he had offered his services to do mathematical calculations. The Oort Archives contain a letter dated 25 August, in which Oort was informed that the company would make use of his services if necessary; apparently he had already done some work because with another letter of the same date he received a cheque for forty guilders (now about 270€). Oort used his mother's address in Oegstgeest for this. From the Archives we also learn that on November 3, 1942 he was dismissed from the position of deputy director, but was given leave for ten months to write a book about the dynamics of our Galaxy. On 12 December, the Minister gave him permission to change his domicile, so that no longer an amount of his salary was withheld for living in the official residence at the Sterrewacht. There are also two letters dated January 28, 1943, in which the Minister granted him honorable discharge as extraordinary professor and at the same time granted a new appointment as 'managing' deputy director of the Sterrewacht. The salary was 5040 guilders (now about 34,000€ in current purchasing power or then abour 4 times that of an unskilled worker). Oort was warned that this appointment was temporary, but no end date was mentioned. So, it could be extended routinely and this at least solved the problem of an income. The interim arrangement of managing deputy director almost certainly continued until the end of the war.

From the pocket diaries (Wednesday November 4, 1942):

'Leave for Hulshorst. Mieke, Coen and Bram by taxi, Marijke and I with the tandem bicycle to the railway station for the 9 o'clock train. Forget the small bag of Marijke and Bram! Still have 8 pieces of luggage, which we transfer with difficulty in Woerden into, and in Utrecht out of, the train with the help of 2 gentlemen who unload it through the window. In Utrecht, Marijke and Bram stay behind [with their grandparents] and Coen comes with us. After much delay due to derailment in Amersfoort and bombs near Stroe, we arrive 5 hours late in the dark in Hulshorst. Luckily we can eat and sleep at Lies Tinbergen's and go to Den Potbrummel on Thursday morning, with clear cold weather, which has been made most inviting by Mr. and Mrs. van Osselen, with flowers and fruits. We feel right at home. Coen and I drive quite a few times back and forth for luggage and blankets. [...]

Saturday baggage arrives in a sealed wagon. Migchelsen calls us and the truck driver De Zwaan drives everything in 2 carts to the Potbrummel, where the whole thing is transferred into the house before noon. Coen does not feel well and stays bed all day. I am lugging and opening the wooden cases. We had no less than 19 cases + 10 bags of anthracite and the piano.'

Nikolaas (Niko) Tinbergen (1907–1988), the later Nobel Prize winner in physiology or medicine, was a biologist, who had been appointed as a lecturer in Leiden in 1940. He had grown up in Den Haag, where his father taught Dutch at an HBS. The father rented a house in Hulshorst every year for his family of five children, including Jan Tinbergen (1903–1994), who studied mathematics and physics and also won a Nobel Prize (in 1969 for economics). Later Niko Tinbergen every summer came to Hulshorst, where he organized summer camps for his biology students. He was taken hostage and interned in Sint-Michielsgestel in September 1942. His wife Elisabeth (Lies) Rutten (1912–1990) and their son Jaap (1934–2010) lived in Vierhouten, less than 10 km from Hulsthorst. Niko Tinbergen was released on September 3, 1944. Jaap Tinbergen would later become an astronomer.

The Potbrummel (see Figs. 7.10 and 7.11) was not luxurious; there was no heating, electricity or running water. From Marijke Oort's speech at the funeral of her brother Coen (November 29, 2006), made available to me by the other brother Bram Oort:

'My clearest memories date back to wartime, when Coen and I were very dependent on each other. Together we had to travel on a heavy tandem [bicycle for two persons] from Hulshorst to our secondary school in Harderwijk, which was almost an hour by bike. If you had to prepare for an exam, you would sit

Fig. 7.10 The Oort family outside the Potbrummel. Photograph made available by M.J.A. (Marc) Oort.

Fig. 7.11 Another picture of the Potbrummel with Mrs. Oort, Marijke and Bram on the terrace. The photograph is dated September 1943. Photo made available by M.J.A. (Marc) Oort.

in the back with the book between the hockey stick clamps and maybe learn something more. When it was my turn (we studied by speaking out everything loud, of course) Coen would turn his head every time and shout 'ridiculous', meaning I had to learn completely useless things. He also regularly shouted to the back: 'You're not pedalling again'.

Often, when the English planes fired on the road, we had to dive into the ditch with the tandem. After a year our parents kept us at home, because it had become too dangerous. We always were sitting around the pot stove. In the evenings, we had to take turns to work the bicycle that stood in the room to provided electricity and light for the family's activities.'

And from Bram Oort's own notes:

'Nobody had electricity so we used a stationary bike to pump water from a well and to have light at night. My mother cooked simple meals on a wood-burning stove; I remember that at crucial moments my older brother stopped pedalling, causing the milk to boil over and spread a nauseating odor. Several of my father's colleague's were also in hiding in this area. They organized seminars to keep the mind sharp. I remember sitting at the back of my father's bike, going over bumpy dirt roads through the pine forests. Usually there were no Germans to be seen on those paths. Often we came back in the dark, which was officially forbidden. New tires were not available – some people biked on noisy, solid tires. We often had to stop to fix a flat tire; the tubes were completely covered with stickers.'

Duyvendak was arrested in March 1943, but released again in August; he also left Leiden and found a plac to hide in Ermelo, only 15 km from Hulshorst. He, Oort and Tinbergen (after they were released in September 1944) regularly saw each other. Also in September 1944 Oort's brother John and his family joined them. Their house in Wageningen was so damaged during the battle of Arnhem that it was no longer habitable. They moved into the Potbrummel and Oort and his family were allocated a part of Groeneveld, the larger and more comfortable home of the van Osselen family.
From Oort's diary entries:

'Monday afternoon, November 13th [1944]. On Groeneveld, where we live from October 3rd onward, while Jo and Pivy now are in the Potbrummel where we fortunately are still a little bit at home. [...]
December 26th (Boxing Day 1944). [...] Then there are these pressing concerns about food. In the cities of Holland and Utrecht, the bread ration has been reduced to 1000 grams per week, the potato ration to 1 kg, so that people are already suffering from hunger. We still have 1400 grams of bread and 3 kg of potatoes, but the last few weeks no more butter and an uncertain future in front of us. So last week we joined the endless procession of bikers and pedestrians, with or without handcarts, who are moving along the road to the North and returning fully loaded very often walking or pulling. [...] It is a sad sight to see all those people going out on the bonne foi [on the off-chance], mostly with their children with them, hoping they will find something to eat along the way. For example, I saw a man with a girl of less than 14 years old stumbling behind a handcart. They came from Meppel where they first had walked to from Utrecht to find some potatoes. Girl was almost exhausted and only had rags around her feet. You see women with their feet wrapped in rags and children by themselves. One night there were two of them on Groeneveld, 14 and 10, who had been on their way from Hilversum with a cart for 5 days and who had exchanged 2 dogs they had brought with them for about 10 kg of potatoes. [...]
Mieke and I beg for a day in Doornspijk near Oldebroek, Marijke and I the next day, but come home with flat tires, which can not be repaired at a bicycle shop, because they are out of adhesive and have no time. Many cycle on solid tires and even more just on the rims of the wheels. And that with the heaviest loads possible. The other day Coen and I cycled for about 45 km to Genne and the Overijsselsche Vecht to get potatoes. On the way back in the dark between Wezep and Oldebroek we had a ruptured tire. Spent the night with a farmer, like all these 'trekkers' do when they can. Many that can no longer find shelter are being accommodated in schools. But there is nothing to eat for them there. The attitude of the farmers is generally admirable. They let the townspeople into their homes and feed them. In case they sell food they hardly ask more than the official government price for their rye and do not want to take advantage of their countrymen.

28 Jan [1945]. The situation gets more and more precarious. In addition to the hunger that forces people onto the roads (as a result of the frost there the 1 kg of potatoes per week now has also been canceled some weeks ago in most cities; even here the last 14 days no potato vouchers have arrived and no potatoes are available. The central kitchen [providing food for people that have no other means to be fed] provides only some rather thin soup, mixed with some carrot or beets). In the last month an increasingly sharper cold has arrived and in addition snow that makes travel extremely difficult. In spite of that the 'trek' has increased enormously. Two independent counts found that 15000 and 16000 people pass by us each day in both directions. The road is completely full so that you have trouble all the time overtaking others and sometimes in villages the road is completely blocked. The sight of returning people, of which half are on foot, are a dismal sight. Often they fall on the slippery roads and can only get back up with the help of others who also help to pick up the bicycle.

It's surprising that so few have died. Dr. Zijlstra told me of 2 deaths in the last two weeks, a woman of 45 who almost suddenly dropped dead and a child of 14 who was travelling with his brothers of 17 and 16, but were now returning home to their parents with the corpse on their handcart.

10 Feb. '45. These days, the bread ration in Holland and Utrecht was temporarily reduced to 500 g (as a result of the frost). But now again 800 g per person per week. The shortages are becoming hopeless, although here and there they are supplemented by sugar beets and in Leiden mainly by tulip bulbs, which are eaten on a large scale instead of potatoes. They are very tasty, reminiscent of sweet chestnuts and have a greater nutritional value than potatoes. These days, when I was on a 2-day visit to Leiden, I ate them in all kinds of dishes. They have caused a real revival in Leiden, are however quite expensive (I heard f. 0.75 per KG [about the hourly wage of a worker]) and can now be obtained only with difficulty. [...]

Feb. 45. After a new trip to the North, with twofold goal: first to get the 15 'mud' potatoes [a 'mud' is a volume unit of 100 liters, which with potatoes amounts to some 70 kg] that Jo had ordered in Dedemsvaart for both our families and that of v.d.Brink, to Hattemerbroek and secondly to try to buy 25 'mud' for the staff of the Sterrewacht and have it shipped at Dedemsvaart. The transport of the potatoes [...] is much more difficult than we expected. More than 5 'mud' at a time is impossible to lug up the slope along the road to the bridge and the control post and even that with local help of others. The German soldiers provide assistance apparently to relieve their conscience, for they help diligently with pushing when necessary.

March 3rd. The last two days of February I went with Duyvendak on an expedition to Zwolle to bring potatoes he and the Tinbergens have bought across the bridge. Strange experience: the calm with which Duyvendak deals with all of this. He pedals very slowly because he has recently been ill, and does everything else with a quiet serenity worthy of a real son of the Heavenly Kingdom. All the time I am nervous, thinking about how short the time available will be before

closure of the [river] IJssel, and how impossible it seems to get the current 20
mud still across the bridge.

March 5th. In a few days, after a crowd that was worse than ever before, calm
has returned to the streets, now that the entire IJssel is closed since March 1st.
Just a single handcart or biking 'trekker' is passing by, where not long ago an
uninterrupted procession went by.'

From a loose leaf 'About the adventures of astronomers in Leiden':

'In assistant's house 'Trouw' [an illegal newspaper] for the last 2 years. After
'Dolle Dinsdag' this newspaper was stencilled in 2500 copies at Sterrewacht 2 for
Leiden and surroundings (once every 14 days) and distributed from there, with
Steinmetz working hard. When the Allies arrived in the south of the country he
wanted to join them, but was arrested in Antwerp. Hins plays an important role
as reserve officer [in the Army before the war], kept weapons hidden in his own
basement.

Dolle Dinsdag (Mad Tuesday) was September 5, 1944, when false reports
were spread that the Allies would liberate Holland at any moment, so that
many started celebrating in the streets. Carl Hermann Dino Steinmetz (1920-
<1999) studied astronomy in Leiden, but dropped out in 1947. In the Oort
Archives there is a letter from him to Oort dated in 1958, in which he informed
him that he now was a professor of physics in Medan (Sumatra, Indonesia)
and would soon get a permanent appointment. But a year later he wrote
that because of the conflict between the Netherlands and Indonesia over New
Guinea he had to leave the country and asked Oort to write a reference letter
for a position at UNESCO. After that, there is no further correspondence with
him.

Several employees of the Sterrewacht had a share in resistance work. Herman
Kleibrink described his activities, such as maintenance of radio equipment
and the like, in an interesting interview [88]. In it he also mentioned Pieter
Oosterhoff as having been active in the resistance. People often for security
reasons did not even know that about each other.

7.6 Interstellar Medium

Oort was supposed to use his time in Hulsthorst to write a book about the
structure and dynamics of our and other galaxies. He indeed worked on this
and a part of his publications up to and including 1942 can also be seen
as preparations for this by taking stock of a certain subject, often related to
astrometry and stellar positions. He had already had some correspondence

about this with Gerard Kuiper and Pieter van Rhijn in 1941. The Archives also contain drafts of a table of contents (the title of the book would be *The Galactic System*). Of the first chapters, 'general introduction' and 'types of observations', there exist detailed draft texts, but of the other chapters nothing but titles. After the nearby stars, star clusters, motions of stars and the interstellar medium, Oort planned to deal with the structure of the stellar system and finally its dynamics.

Now that it was clear that dust is widely distributed between the stars in interstellar space, which results in a strong absorption in the plane of the Galaxy, Oort wondered what the properties of that dust were and also how it was formed. Even after he moved to Hulshorst, Oort still regularly came to Leiden. He also secretly gave lectures there, in the darkroom under the photographic telescope. Hertzsprung knew nothing about this; when he knocked on the door of the darkroom he was told that whey were developing photographic plates and no light was allowed so they could not open the door. Before he moved to Hulshorst Oort already had had contacts with Hendrik Anthony (Hans) Kramers (1894–1952), physics professor in Leiden, on the question of the formation of molecules and smoke particles in the interstellar medium. Kramer and Oort had organized an essay competition (which happened with

Fig. 7.12 Henk van de Hulst and Oort during the first Dutch Astronomers Conference in 1941. From the Dutch Astronomers Club [86]

some regularity), under the auspices of the Rector Magnificus and the Senate of the university. The subject was the question how and to what extent over a period of billion years, the order of the age of the Solar System, solid particles can form in interstellar gas clouds. And what would be the composition and distribution of dimensions of those particles. The deadline was May 1, 1942. One of the entries came from from a student of Kramers, Dirk ter Haar (1919–2002), who was working on a dissertation on the origin of the planetary system under Kramers. A second entry came from a student in Utrecht of Minnaert, Hendrik Christoffel (Henk) van de Hulst (1918–2000). He and Oort had met at the astronomers' conferences (see Fig. 7.12). Minnaert had encouraged van de Hulst to participate. The third entry came from Adriaan van Woerkom.

Van de Hulst and ter Haar received an honorable mention; there was no winner. Minnaert arranged that Henk van de Hulst (see Fig. 7.13) could work in Leiden for some time, where he attended lectures by Oort and continued to work on the subject. Later, when he was back in Utrecht, Oort looked him up stopping by when on his bicycle he was on his way to Leiden. It eventually

Fig. 7.13 Hendrik Christoffel (Henk) van de Hulst as a young man. From the Archives the Sterrewacht Leiden

Fig. 7.14 The antenna that Grote Reber had built in his backyard to study 'cosmic static'. The picture which Oort must have seen in Reber's 1940 article is less clear than this one, which comes from a publication in 1944 [89]. The surface is made of metal and the diameter is 9 meter. Oort probably saw this picture only after the war. From Wikicommons [90]

resulted in an article by Oort and van de Hulst, published after the war, in which they made a study of the growth and composition of those dust particles. They found that the density should be 1.3×10^{13} particles per cubic centimeter, that they should have an average size of 0.15 microns and would exist for an average 5 million years. Van de Hulst later in his career made a big name for himself with studies of the scattering of light on interstellar particles.

Occasionally there were still meetings, known as interacademic colloquia, of the Dutch Astronomers Club, at which recent literature was being discussed. Oort had seen two articles in which radio radiation from the Universe was discovered. The first was from 1932 by Karl Guthe Jansky (1905–1950), who

Fig. 7.15 Reconstruction in 1953 of the colloquium at the Sterrewacht Leiden originally held in 1944. In the first row Houtgast, Bakker en Oort. From the Archives of the Sterrewacht Leiden

investigated the cause of radio telecommunication interference for the Bell Company. He found that part of the 'static' came from the Galaxy. Grote Reber (1911–2002) had built a radio telescope in his hometown Wheaton, Illinois, with his own hands (Fig. 7.14), with which he had better mapped the radiation of the Milky Way, and had published an article about it in the astronomical literature in 1940. Oort suggested to take this as the subject of one of those meetings, and that Henk van de Hulst would discuss the issue of the mechanisms by which radio radiation is emitted. What Oort did not know then is that Reber had published a second article in 1944.

Actually, there was only one known process. Star formation creates a whole range of masses. The heaviest stars become very bright and live short (see appendix A). Such stars, which have spectral types O or B, can therefore only be found in areas of recent star formation. Now they are very hot and emit large amounts of ultraviolet light. That light is able to remove electrons from atoms in the gas around it, like of hydrogen, which has one electron (and a nucleus containing one proton). So the area around areas of young stars has been ionized by this light. Now two things can happen when an electron encounters a proton (or hydrogen nucleus). They can form a hydrogen *atom* again, in this process emitting a photon containing the excess energy available. This is called recombination. Usually that electron first ends up in a high energy

Fig. 7.16 Photograph of the 1953 reconstruction of the colloquium at the Sterrewacht in 1944 from a different angle. From the Archives of the Sterrewacht Leiden

level and then eventually falls back via lower levels to the ground level. Around hot, young stars we find areas that for the most part provide light in the spectral lines resulting from this process of falling back to lower energy levels. But it is also possible that the electron is not captured, but only deflected in its orbit. No atom is created then, but the electron loses energy and radiation is emitted. This is called free-free radiation and corresponds to electromagnetic radiation in the radio domain. Van de Hulst would discuss this process in the first place. This radiation is not restricted to specific wavelengths (like a spectral line) but is emitted at all possible wavelengths, i.e. it is continuum radiation. But Oort also asked van de Hulst to investigate whether there could be a spectral line in the radio domain. If so, one could measure the wavelength and use the Doppler effect to measure the velocity of the gas that that line had emitted. This would provide opportunities to study the Galaxy in areas where absorption makes it impossible in optical light.

The colloquium was held at the Sterrewacht Leiden on April 13, 1944 (Figs. 7.15 and 7.16). For technical background Oort had asked a physicist from Philips NatLab to be the first person to speak at the colloquium. This was Cornelis Jan Bakker (1904–1960) who spoke about the possibility of developing sensitive receivers to study cosmic radio radiation. Van de Hulst spoke second and spent most of his time on the continuum radiation and only at the end presented calculations about a line of neutral hydrogen with a radio wavelength, which he predicted might be observable.

A spectral line in the radio area must be related to a transition in an atom, which corresponds to relatively little energy. The hydrogen atom consists of two particles, a proton and an electron, both of which have a property with the character of rotation; these are called the spins. Now, according to quantum mechanics, the spins of the proton and the electron can only be parallel or anti-parallel, and there is a small difference in energy the electron has. All levels have this property, but most hydrogen in interstellar space is in the ground state. In the lower of these two energy states within the ground state, the spins of proton and electron are anti-parallel, but part of the hydrogen will be in the slightly higher state, where they are parallel. It takes a long time, on average ten million years (it is a so called 'forbidden' transition, because in a laboratory an atom is more likely to first collide with another particle), but then the electron falls to its lower state and a photon is emitted. This has a frequency of 1420.4 MHz, or a wavelength of 21.1 cm.

Oort decided to build a radio telescope and start looking for that line as soon as the war would be over.

Fig. 8.0 Oort in 1946. From the Oort Archives

8

Breaking New Ground

It is through science that we prove,
but through intuition that we discover.
Jules Henri Poincaré

Maybe some people don't feel scared
when they think about comets and supernovas.
Maybe they think it is wonderful.
Lydia M. Netzer (b. 1972) [91].

The Second World War was justifiably a watershed. Until its outbreak, Oort's scientific work had almost exclusively been within the field of astrometry and the statistical astronomy of Kapteyn, with the aim of studying the structure and dynamics of our Galaxy, and of the distribution of starlight in extragalactic systems as a first step towards the dynamics of those galaxies. Just before the war there were two exceptions of research in other astronomical fields. The first was still in the tradition of his great teacher Kapteyn, where it concerned the study Nova Persei 1901, in which apparent expansions were now seen faster than those of light. The second was related to the identification of the Crab Nebula with a supernova from 1054. Both have been discussed in the previous chapter.

The watershed was, as Oort himself formulated it, the result of the rise of radio astronomy, which allowed a new view of our Galaxy, unhindered by interstellar dust, and allowed the spiral structure to be studied. Oort's intuition

---

Lydia M. Netzer, née Jenkins, is an American author. From her book *How to tell Toledo from the night sky.*

P. C. van der Kruit, *Master of Galactic Astronomy: A Biography of Jan Hendrik Oort,*
Springer Biographies, https://doi.org/10.1007/978-3-030-55548-1_8

proved to be right; it opened up a whole new field of research with surprising new insights. Oort also took other paths and entered other unexplored terrains. He continued research with Henk van de Hulst on the formation of dust particles in space, and with Lyman Spitzer of Princeton he studied the question of how interstellar gas clouds obtain and maintain their velocities. And his fascination expanded to the question where comets come from and how remnants of supernovae produce their radiation. Comets and supernovae, traditionally heralds of disaster.

8.1   Director of Sterrewacht Leiden

The end of the Second World War took place in the Netherlands on Saturday, May 5, 1945. Already on Monday, May 7, Oort went with Nico Tinbergen to Zwolle 'to get papers for a trip to Holland. Lovely weather. Bridges have been destroyed near Zwolle'. Apparently this was successful and on May 10th (Ascension Day) Oort, Duyvendak and Tinbergen left on a tandem bike and a bicycle for Leiden. They stopped in De Bilt, where Oort's in-laws lived. Oort had already noticed from afar that the flag was not flying. It turned out that on April 18 his father-in-law Willem Graadt van Roggen had died. The funeral obviously had taken place without them being informed. Duyvendak and Tinbergen continued the journey to Leiden and Oort talked for a long time with his mother-in-law and then cycled back to Hulshorst. He wrote about this in his diary:

> 'When I arrived, the three children were outside, all dressed up in orange clothing, ready to go after a children's party in the afternoon, to the tour along the Vierhouten cycling path. When they saw me, all three of them rushed towards me in joy. Oh, their poor sad faces when I said there was something very sorrowful, and went up to Mieke. Later I bring them upstairs and we tell them. For a while we sit next to each other on Bram's bed, belonging together very intensely. Bram crying so pathetically, Mieke so terribly sad, Coen fighting back his tears without saying anything, but so pale. Marijke later said that she had known when I went upstairs. We read Grandma's letter and Mieke is so sweet. Marijke was sobbing when we told her: 'Grandmother has to come here.'

It turned out that at the funeral Oort's mother-in-law had given a short speech and then had asked to sing the Wilhelmus, the national anthem. This of course was strictly forbidden. Apparently they got away with it.

On Tuesday May 15, Oort left again and slept in De Bilt. The next day he went on to Leiden (according to his pocket diary he left at half past one and arrived in Oegstgeest at about 7 o'clock, that is 65 km in 4.5 h). A few days later, after having spoken several people, he returned to Hulshorst. On their wedding day, May 24th, he presented his wife with roses from the terrace of the Sterrewacht, which are 'still beautiful and fragrant'.

On June 5, Oort and Mrs. Oort left together for De Bilt. Oort went on by himself to Leiden, and after a few days they returned separately. In order to also go to Leiden, Mrs. Oort had to have a pass to cross the Grebbe line, a more or less north-south running defense line established in 1725 and built for inundation, where not all mines had been cleared yet. That succeeded with some delay, and on June 18 they were ready to move back to Leiden. Oort travelled back by bike, and the rest of the family apparently independently. In his pocket diary Oort also wrote that on that day he had been appointed manager of the Sterrewacht. And on 19 June: 'First day as 'boss' at the Sterrewacht, where for the time being work is only done during half days'. On Wednesday June 20, he gave a short speech to the staff during morning coffee, 'surrogate, arranged by Mieke, with sugar sent by van Hoof [Armand van Hoof, astronomer of Leuven, Belgium (1906–1989)] and there are also cigarettes from Kuiper!'

Gerard Kuiper was by now a naturalized American. During the war he had taken leave of Yerkes Observatory to do radar research at Harvard's Radio Research Laboratory. In 1945, he had come to Europe as part of a mission, which sought information about the German development of atomic, biological and chemical weapons. In a daring action in the Russian area, he had kidnapped the famous physicist Max Karl Ernst Ludwig Planck (1858–1947) to Allied territory with men under his command, hours before the Russians arrived. He was probably visiting Leiden to see how things stood there.

Oort's interim appointment as manager of the Sterrewacht had apparently been tacitly continued. It took until August 28 before the appointment became permanent again and until November 8 before he was appointed professor of astronomy and director of the Sterrewacht. Hertzsprung had remained director during the war, even after his seventeenth birthday in 1943. In September 1945, he returned to his native Denmark. He only came back to Leiden for a formal farewell on June 25, 1946. During this ceremony Oort compared him in his speech with the other famous Danish astronomer Tycho Brahe, in which he praised both of them for their attitude that real progress can only be made with accurate observations.

There were no deaths among the staff of the Sterrewacht. There were, however, some among former students and employees who happened to be in the Dutch East Indies when the Japanese occupation began. In 1923 an observatory was set up in Lembang near Bandung by Joan George Erardus Gijsbertus Voûte (1879–1963) with the financial support of tea planters and amateur astronomers Karel Albert Rudolf Bosscha (1865–1928) and his cousin Rudolf

Eduard Kerkhoven (1848–1918). These two gentlemen are the main characters in the well-known novel by Dutch writer Hélène Serafia (Hella) Haasse (1918–2011), that tells their story [92]. Aernout de Sitter had moved from Johannesburg to Lembang in 1939 to become director of this Bosscha Observatory. Another PhD student of Hertzsprung, Willem Christiaan Martin (1910–1945), had been appointed there in 1940 also after having worked in Johannesburg. They, and Jan Uitterdijk (1907–1945), who had completed his astronomy studies in Leiden and had become a teacher in the Dutch East Indies, all died in Japanese camps.

De Sitter's wife, Henriëtte Johanna (Hette) Zoetelief Tromp (1906–19??), and their four children survived the war and returned to the Netherlands, where the Oorts took care of them as if they were in the family. Oort published in *Hemel & Dampkring* a commemoration of Aernout de Sitter [93]. Some time later, in January 1946, lecturer Jan Woltjer (1891–1946) died due to lack of food and other hardships during the war.

8.2 Recovery of the IAU

During the war, Oort had been General Secretary of the International Astronomical Union IAU. In the first months of the war, he had been able to transfer current affairs to one of the Vice-Presidents, Walter Adams, director of the Mount Wilson Observatory. The IAU had bank accounts in dollars, pounds and guilders; the latter account could of course not be used for international transfers during the war, but was kept alive. IAU President Arthur Eddington had died in 1944. Oort had to take swift action to bring the union back to life. With the help of Kuiper, Oort flew to London on September 30, 1945 and spent several days in the U.K. to consult with his British colleagues. Fortunately, Adams had arranged for former vice-president Harold Spencer Jones (1890–1960), Astronomer Royal, to take over the presidency so that it remained in the United Kingdom.

Oort and Spencer Jones organized a meeting in March 1946 in Copenhagen, at which the IAU Executive Committee met with leading astronomers from most countries. Although some, such as Walter Adams, seem to have felt uncomfortable about it, this time no measures were taken to exclude astronomers from the defeated nations. Spencer Jones and Oort were reconfirmed as President and General Secretary and instructed to organize the next General Assembly, to be held in 1948 in Zürich, Switzerland, in (former) neutral territory. Individual astronomers from Germany and Japan were welcome; the assumption was that it had been established by the relevant authorities that they had not committed any war crimes. Of the 279 participants in Zürich,

Fig. 8.1 Change of the guards at the 1948 IAU General Assembly in Zürich. The incumbent President Harold Spencer Jones and General Secretary Oort (second and third from the left) were succeeded by Bertil Lindblad (left) and Bengt Strömgren respectively. From the Oort Archives

eight were indeed German, but there were not Japanese. Oort was relieved of his duties as General Secretary, and in gratitude for his efforts over a long and difficult period of time, he received a standing ovation. It should not go unmentioned that Heleen Kluyver had taken a large part of the work out of his hands. Bertil Lindblad was elected the next President and the Swede Bengt Georg Daniel Strömgren (1908–1987), the son of Svante Strömgren, General Secretary (Fig. 8.1).

8.3 The Interstellar Medium

During his lifetime Oort has received many important honors. Already in 1937 he was appointed member of the Royal Netherlands Academy of Arts and Sciences (KNAW). In January 1942 he received a letter, dated December 4, 1941, from the Astronomical Society of the Pacific, awarding him the prestigious Bruce Medal. This was named after Catherine Wolfe Bruce (1816–1900), a generous donor of funds to, among others, astronomy. The first Bruce Medal was awarded to Simon Newcomb in 1898 and the third in 1900 to David Gill. Kapteyn had received the Bruce Medal in 1913, Willem de Sitter in 1931 and Ejnar Hertzsprung in 1937. On the letter, Oort wrote in pencil that the day before the date of the letter, the war between the United States and Japan had broken out (incorrectly, this was on December 8). According to the Society notes, no reply was received from Oort. After 1941, no more medals were awarded until the end of the war. On December 31, 1945, the Secretary of the Society informed Oort of the fact that the medal had been forwarded via the Dutch consul in San Francisco to Professor Pannekoek in Amsterdam with the request to hand it over to Oort at an appropriate opportunity. Pannekoek did this at the gathering of the Dutch Astronomers Club on May 4, 1946, where he read a letter from the President of the Society, Arthur Scott King (1876–1957) of Mount Wilson Observatory [94].

Meanwhile, on January 12, 1946, Oort had received a telegram that the Royal Astronomical Society had awarded him the Gold Medal. Here Oort was preceded by Kapteyn in 1902, Hertzsprung in 1929 and de Sitter in 1931. The ceremony associated with the award of the Gold Medal had been set for May 10, 1946, so after receiving the Bruce Medal, Oort had to travel quickly to London for the Gold Medal. This ceremony was preceded by the 'Darwin Lecture', a lecture named after George Howard Darwin (1845–1912), fifth child and second son of the famous Charles Robert Darwin (1809–1882). George Darwin was a lawyer, but later became professor of astronomy at Cambridge. His work mainly concerned the interaction between the Earth and the Moon and the formation of the latter.

The subject of Oort's lecture in London was *Some phenomena concerning the interstellar medium*. The text was published in [95]. Where Oort was of course honored for his work on the dynamics of the Milky Way, his interests had shifted to the gas and dust between the stars. Oort began to notice that this medium is very unevenly distributed throughout space. He then summarized the work of Henk van de Hulst and himself on the formation and distribution of sizes of particles in interstellar space. He then treated the distribution of dust in other galaxies. In some galaxies there is no indication of dust, such as NGC 3115 (Fig. 6.19). His analysis presented at the inauguration of the McDonald Observatory in 1940 had shown that there had to be a lot of matter that gave very little light. Oort speculated that the gas and dust had been exhausted by the formation of stars, and that this now had to be mainly in the form of light stars that emit little light. He then brought up the dust-lane in NGC 4595, the Sombrero Nebula (Fig. 6.5). This would, according to Oort, have the shape of a ring; in the inner parts star formation would have used up the gas, in which spiral structure might have been an important factor. The

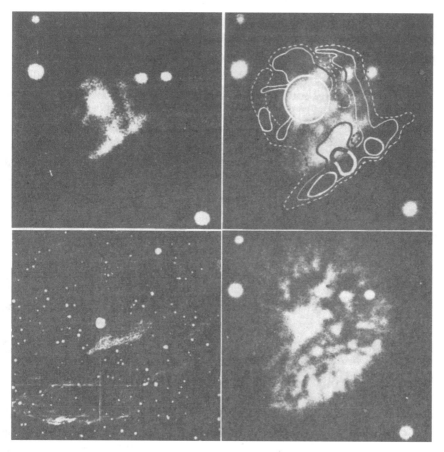

Fig. 8.2 Nova Persei 1901. Above: November 15, 1917 (left) and October 3, 1934. Bottom: November 13, 1901 (left) and September 28, 1943 (right). The lines on the upper-right represent the distribution on the lower right. The picture at the bottom-left is a drawing from a photograph. This is an adapted version of a figure of Oort in his Darwin Lecture. From the Oort Archives

formation and maintenance of spiral structure would remain an important theme for Oort.

The main theme came after that, and that was the question of collisions between gas clouds and the interaction of expanding shells of novae and supernovae with the interstellar medium. Apparently that interested him the most at that time. Where does the light emanating from such shells come from, as for example in the expanding nebula around Nova Persei (Fig. 8.2)? Oort estimated that the energy emitted was about equal to the energy that was lost by the interstellar gas slowing down the expanding shell, and that the light of the nova was insufficient to provide that. Oort reasoned that in the collision of the expanding shell with the gas present, atoms would be ionized, releasing electrons. Electrons lose energy by passing the atoms present, so that their average energy decreases. In areas such as around hot stars, where ionization by starlight takes place, the energy is then limited to an equilibrium value, the energy per electron corresponding to a temperature of some 10,000 K. But in the colliding shells the temperature of the electrons only increases because of the energy in the collision, so that when they hit an atom those electrons will end up in higher energy levels in the recombined atom. In oxygen, and other atoms also, there are transitions in such higher levels that do not occur in the laboratory. The average time that an electron stays in that level before such a transition to a lower level takes place in the laboratory is too long, because the atom will already have collided with another atom (taking the electron out of the level) before the transition can occur spontaneously. But in the interstellar medium these collisions are much rarer and these so-called 'forbidden' transitions will occur. According to Oort, this explained why these forbidden transitions are so strong in the shells and much less in HII regions (regions of ionized hydrogen) around hot stars. This interpretation turned out to be correct.

Oort mentioned a few other examples, such as the Veil Nebula and Cygnus Loop (see Fig. 8.3). Sometimes the Veil Nebula (or Cirrus Nebula) only refers to the right-hand side, sometimes to the entire structure. The question was why the nebula is so thin. Oort suggested that this was due to strong compression of the interstellar gas by the shell and that this would mean, as the Swiss-American astronomer Fritz Zwicky (1898–1974) had already suggested, that it would be a supernova remnant. The expansion on the sky was 0.03 arcseconds a year, and that meant an age of 6000 to 7000 years. The expansion rate was measured spectroscopically (about 75 km/s); the distance would then be 500 pc and the radius 10 pc. Nowadays this is estimated to be three times as much.

Finally, Oort pointed to the Crab Nebula (see Figs. 7.4 and 8.4). Here the mass of the ejected shell was too large and the time scale (since the supernova in 1054) too short to have caused much deceleration. The relationship between the amorphous structure and the surrounding filaments of ionized gas was unclear. The Crab remained an intriguing object and Oort would do much more research on it, as we will see later.

8.4   Yerkes, McDonald and Palomar

In 1947 Oort decided that it was possible for him go to the United States again for some time. This was possible because Pieter Oosterhoff, as deputy director, had taken on many tasks that concerned the running of the Sterrewacht. When he went to the opening of the McDonald Observatory in 1939, Oort had had

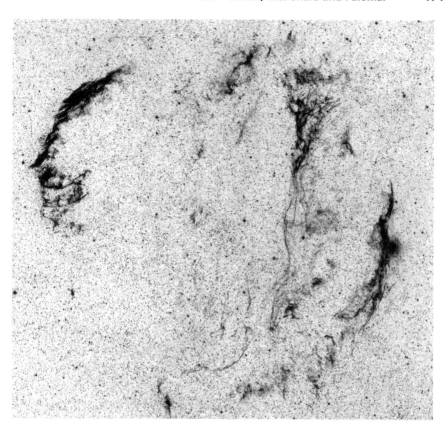

Fig. 8.3 Cygnus Loop or Veil Nebula in the constellation Cygnus (Swan). It is the remnant of a supernova explosion. The dimensions in the sky are large; this picture covers an area of about 2.7 by 2.5 degrees in the sky. The photo was scanned from a red print of the National Geographic Society—Palomar Observatory Sky Survey

an invitation to spend some time at Yerkes Observatory and to lecture (for students of the University of Chicago), but had turned it down in order to be able to go observing on Mount Wilson. There he had finally succeeded in obtaining a large number of photographic plates on galaxies in order to map the light distribution. The processing progressed, work that was again done by Herman Kleibrink. We already saw that the technique was to measure the blackening of the photographic emulsion on the galaxy and compare it with positions on the sky only just next to the galaxy. In order to relate the blackening to the amount of light that had caused it, there were also exposures made of stars of known brightness in Kapteyn's *Selected Areas*. These were taken out of focus so as not to overexpose the plate. There was a second possibility, namely to expose the plate at the edge with a set of 'spots', of which the relative ratio

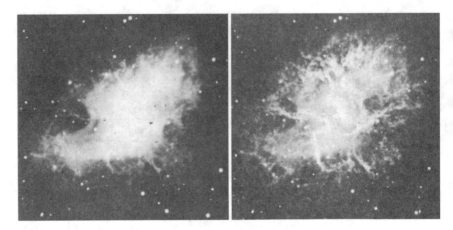

Fig. 8.4 Photographs of the Crab Nebula by Walter Baade. Left in blue light, right in that of a spectral line of hydrogen, Hα, which corresponds to the transition from the second lowest energy level to the one but lowest. Adapted version of a figure from Oorts Darwin Lecture. From the Oort Archives

of the brightness during exposure was known. This of course provides only a relative calibration, but it would be an independent check on the relative magnitude scale. The problem was that these two calibrations turned out to be inconsistent with each other.

Otto Struve had renewed the invitation to come to Yerkes and Oort decided to accept the invitation this time. Struve, before Oort came for the visit, had in the mean time become director of the astronomy department of the University of California at Berkeley (near San Francisco) in 1947 just before Oort's visit and Gerard Kuiper was now director of Yerkes. But Oort had a second reason and that was to do observations designed to solve this inconsistency. William Albert Hiltner (1914–1991), who worked at Yerkes, had started upon Struve's recommendation an innovative program to measure the brightnesses of stars using a photoelectric cell. That was much more direct and accurate than from photographic plates. Oort wondered whether this technique could also be used on extensive objects such as galaxies. Hiltner was enthusiastic about this and observing time was obtained on the 82 in. Telescope of McDonald Observatory in Texas to try it out.

Oort left on September 13, 1947 and stayed as always after arrival in New York a few days with Jan and Jo Schilt, who meanwhile lived in New Jersey. Oort arrived in Chicago on October 2. He lectured extensively there and talked to colleagues and after that, in early December, he first traveled to California to go on later to Texas and the McDonald Observatory in January. On December

11, he arrived in Pasadena, where he had many conversations with Edwin Hubble, Walter Baade and Milton Humason.

Oort had a very special experience. On Palomar Mountain, about 150 km southeast of Pasadena, the then largest telescope in the world was built. After the 60 in. and 100 in. of Mount Wilson, the 200 in. (5 m) telescope was to be George Hale's last tour de force. The project was funded by the Rockefeller Foundation and construction began as early as 1936. As the light from Los Angeles began to be unacceptably bright, Mount Wilson was no longer suitable and the astronomers moved to a place far away from major cities (San Diego is still 80 km from Palomar Mountain). The dome and telescope were already completed in 1939, but the mirror was, partly due to the Second World War, very much delayed. In the end it took eleven years before the mirror was ground into the right shape. In November 1947 it had finally arrived at the mountain. Hubble and Humason took Oort with them to Palomar, where the mirror had just been hoisted into the telescope. It had not yet been aluminized, but the glass was of course reflective itself. The mirror would be aluminized only after a year of testing; it was finally put into use for scientific research in January 1949. From Oorts diary notes:

'December 22nd. Trip to Palomar with Hubble and Humason, where I was one of the first to see star images from the 200 in. mirror together with Hubble. Yesterday night the first tests: mirror very smooth but astigmatism (2 lines as foci 2 millimeters from each other). This was fixed tonight and the mirror looked as good as for the 100″ on many occasions! Knife-edge test large magnification on a stellar images. Stars hover in the focal plane of the mirror. Impressive ride with the platform up and then stepping down into the cage, where the three persons could sit'.

The cage is the cylindrical observing station in the primary focus at the top of the telescope (see Fig. 8.5). It has a diameter of about one meter and thus blocks only 4 percent of the 5 meter mirror. It can be reached with a platform that moves upwards along the inside of the dome, just next to the slit, a gap that can be opened in the dome. The telescope is at that time pointed towards the zenith and the platform ends in such a way that one can step into the 'cage' from above. This is very impressive indeed, as I also know from my own experience. In the cage is, with some difficulty, room for three. In the situation described above, the telescope probably remained pointed at the zenith. In practice, of course, this is not the case, and then it can be quite uncomfortable, certainly with three people. Without an instrument or plate holder, you could see the stellar images 'floating' in space. Astigmatism is bad imaging due to deviations from the correct shape of the mirror, which would be corrected after the first night by adapting the support of the mirror. The knife-edge is

Fig. 8.5 Palomar Observatory's 200 in. (5 m) Hale Telescope shortly after commissioning. The cylinder hanging at the top is the primary focus observing station ('cage'). The platform that brings the observer to it moves upwards along the other side of the slit opening in the dome (to the right of the picture). The dome is over 40 meter in diameter. From Wikimedia Commons [96]

the method of seeing the deviation of the mirror from the right shape. Such inaccuracies result in scattered light around the stellar image and if you cover the area just next to the star with a sharp surface such as a knife and bring your eye close to it, you can see which part of the mirror is responsible for the poor imaging. The view through the 200 in. Telescope must have been one of Oort's most impressive astronomical experiences.

Oort spent Christmas with the Baade's and then traveled to Texas and the McDonald Observatory. Hiltner had designed and manufactured a construction in which the photometer could move with an electromotor on a kind of sledge in the focal plane, with which the brightness could be measured through a small diaphragm along a line in the focal plane, corresponding to a line through the galaxy. The output of the photoelectric cell behind the photometer was amplified and recorded on a chart recorder (graphic writer), with a pen running over graph paper. This went well and measurements were successfully obtained in nine nebulae. Processing the data and the continuation of this project took years and I will come back to that later.

Oort finally traveled back to New York at the beginning of February and from there home. He returned home on February 16, after having been away for four months.

8.5   Kenya and South-Africa

We saw above that the problem of deflection and refraction in the determination of declinations could be circumvented by going to the equator and that the Sterrewacht had organized such an expedition in the thirties. The result had been disappointing. Firstly, time had been too short to obtain any reliable results, and secondly, there had been problems of hysteresis with the instrument, in which the optical alignment changed when the telescope was reversed depending on the direction in which this was done. Oort, who had inherited the leadership of the expedition from de Sitter, had already started to organize a new expedition before the war (see again [75] for a description of the Kenya expeditions).

The first expedition lasted from July 1931 to May 1933, but the observations themselves were carried out between December 1931 and February 1933, so for only a bit more than a year. Coert Hins and Gijs van Herk had been in Kenya for almost two years during this first expedition. Hins was married (and his marriage survived this assault) and van Herk was single. This time it was decided that the expedition would last three years and that van Herk's (meanwhile) second wife Aartje (Attie) Konijnenburg (1897–1975) and also his two children from his first marriage would come along. The second participant was Willem Haske van Zadelhoff (1904–1964), lieutenant first class in the Navy, who was seconded to the Sterrewacht for this purpose (Fig. 8.6). Apparently he had found this attractive as an amateur astronomer and volunteered to take part in this expedition. He must have heard about it through his mother Maria Ludovica Minnaert (1884–1966), a full cousin of the Utrecht astronomer Marcel Minnaert. Van Zadelhoff would also bring his wife, Henriëtte Adrienne Naber (1906–1980) and his two children. At the departure of the expedition in 1947, the four children were between the ages of eight and fourteen.

Funding again came from a large number of sources; there had already been pledges during the war. The costs were significantly higher this time, but by now the Netherlands Organization for Pure Scientific Research (ZWO) was a new possibility. This organization had been established by the government in 1947 to bring the infrastructure for pure (as opposed to applied) scientific research back up to standard, and although the original contribution was a small part, ZWO eventually financed about half the expedition. The director

Fig. 8.6 Part of a group photograph of the staff of the Sterrewacht in 1947, not long before the second Kenya expedition. In the front Oort, Willem van Zadelhoff and Pieter Oosterhoff. Behind Oort Fjeda Walraven and above him Gijs van Herk. In the left-top corner P. de Haan, who went along to Kenya as carpenter and in the right-top corner Adriaan Blaauw, who went to Kenya also for some time. From the Archives of the Sterrewacht Leiden

of ZWO was Jan Hendrik Bannier (1909–1995); although he and Oort had the same first names, he was addressed as 'Henk'. He was a physicist by training, but had not written a PhD thesis; in 1974 he was honored for his efforts for the Dutch scientific community with an honorary doctorate from the Catholic (now Radboud) University of Nijmegen. Pieter Oosterhoff, who was secretary of the Sterrewacht Fund, (founded to coordinate the financial affairs of the Rockefeller Astrograph), took care of the financial administration.

Fig. 8.7 The horizontal telescope at Timboroa Hill, Kenya, its operation being demonstrated by Gijs van Herk during daytime. From the Oort Archives

The location in Kenya where the expedition set up the observation station was in the same general area as that of the previous expedition, but now 3 km to the northwest. It was called Timboroa Hill, and was located at an altitude of 2880 m and only 36".2 (1.1 km) north of the equator. The same horizontal telescope was used, but given the problems with hysteresis, it had been fixed at 7° above the horizon (see Figs. 8.7 through 8.10, made during Oort's visit to Kenya in 1949). A zenith telescope was also used to determine the exact geographical latitude of the station (Fig. 8.8).

The families of van Zadelhoff and van Herk left for Kenya in August 1947. They were accompanied by P. de Haan of the Sterrewacht, who as a carpenter would oversee the work of erecting the buildings. For personal reasons he returned to the Netherlands before this was finished, in March 1948. The observing was started only in October 1948, more than a year after the departure of the participants. It was the intention to observe all 1193 stars between declinations +60° and −50° from the *Dritter FundamentalKatalog* drawn up at the Astronomisches Rechen-Institut in Heidelberg and published in 1937 (*FK3*). The limitation had to do with positions of columns with marks for the calibration of the azimuth. Another 300 stars from the *GC*, the *Boss General Catalogue*, also known as the *Boss Catalogue*, were observed with the zenith telescope. Instead of reading the scales and jotting down numbers on paper, photographs of the scales were taken, but at first these turned out to be insufficiently sharp. To improve this, Adriaan Blaauw came to Kenya between December 1949 and April 1950 with new lenses for the camera, which turned out a real improvement. After his PhD in 1946, Blaauw had left Groningen and

Fig. 8.8 Oort and van Herk, at the horizontal telescope in Timboroa Hill, Kenya, during Oort's visit there in 1949. From the Oort Archives

his supervisor van Rhijn to settle in Leiden, where he was appointed lecturer at about the same time as Oosterhoff was appointed extraordinary professor in 1948.

In the small community tensions sometimes arose, especially in the beginning between van Herk and van Zadelhoff, but later between the latter and a PhD student W. Derksen from Amsterdam, who worked on cosmic rays with Jacob Clay (1882–1955). These are high-energy particles from the Sun or from other objects in the Universe such as supernovae. The great altitude of the observing location helped to reduce the disturbing effects of collisions with atmospheric molecules. The case became better when this person was replaced by Jan Strackee (1924–2009), who would indeed complete a dissertation on this subject. After three years, in July 1950, the van Zadelhoff family left. The

Fig. 8.9 Oort during his visit to the Kenya expedition in February 1949. From the Oort Archives

original period of his secondment from the Navy had expired some time ago; Willem van Zadelhoff now was four years away from his job in the Navy and wanted to take up his career there. The observations went less well than hoped for and the mission was extended with further funding from ZWO. A student from Leiden, Maarten Schmidt (b. 1929), replaced Zadelhoff. Schmidt had grown up in Groningen and had taken his candidate exam (bachelor) there when Oort offered him to come to Leiden. Schmidt arrived in Kenya in September 1950.

Life in Kenya was not without dangers. There had been a big fire in the bushes and trees, from which the observation station barely escaped. In the rainy season the roads became dangerously slippery. The expedition's truck slipped off the road one day in such conditions, when they were on their way to bring Zadelhoff's son to the train to go to his school in Nairobi. Nobody was injured and to his relief the boy heard the whistling of the departing train in the distance, knowing that the next train would go only in another three days. But in another accident, Maarten Schmidt got off the road, the truck overturned and his ankle got stuck, which was broken in several places.

Fig. 8.10 Some of the buildings at the site of the Kenya expedition; picture probably taken by Oort in February 1949. From the Oort Archives

Eventually he recovered from his shattered ankle. In his letters with Oort, van Herk blamed this partly on his youthful and careless style of driving. As is well-known, Schmidt went on to become the discoverer of the quasars in the 1960s.

In September 1951 the measurements were stopped. Although a further extension had been applied for, ZWO decided to terminate the funding. This against the wishes of Gijs van Herk, who wanted to do another experiment with a specially built instrument for observing the zodiacal light. Zodiac light is scattered sunlight on dust particles in the Solar System, which is therefore close to the orbital plane of the planets (the zodiac). It can be seen at night as a faint glow around the zodiac. Van Herk and his family, and independently Maarten Schmidt, returned to the Netherlands in December 1951. Van Herk and his wife and children had been away for four and a half years!

Oort was in Kenya in 1949. He was on the way back from a visit to South-Africa, which he made in connection with the deteriorating situation with the observation conditions at the Rockefeller Astrograph at the Unie Sterrewag in Johannesburg due to the increasing light pollution of the nearby large city. It was the first time that Oort made an intercontinental journey by plane (in 1945 he had already been to London by plane). In his diary he described the beginning of the journey as follows

'On Dec. 7, '48 I depart by DC6 from Schiphol Airport to Johannesburg, seen off by Mieke, Coen and Marijke. An amazing sensation in the largest and fastest of all aircraft to go up to 5 to 6000 m with cruising speed of 260 miles

Fig. 8.11 Oort on arrival in Johannesburg in December 1948. From left to right: Jan Schilt, Willem van den Bos, Oort and Adriaan Wesselink. From the Oort archives

per hour or 420 km/h, length 30 m, width 35 m. Total weight 40 tons, of which 23 tons for the empty plane, in one hour to reach the Alps and after dusk over the Mediterranean Sea after 4 h to arrive in Tunis in the dark.'

From Schiphol Airport to the Alps (about 650 km) at this speed, including take-off, must have taken somewhat longer than an hour. After a stopover in Nigeria, and maybe even more stops, he arrived in Johannesburg after 28 h since departure. He was met by a group of Dutch people (see Fig. 8.11).

Adriaan Wesselink (1909–1995) was now the observer at the southern station of the Sterrewacht. He had obtained a PhD with Hertzsprung in 1938. He would end up in the United States at Yale University. Willem van den Bos had, as we have already seen, transferred from Leiden observer to the Unie Sterrewag. Schlesinger's Yale telescope was used since 1943 in collaboration with the astronomy department of Columbia University, of which Jan Schilt was now director, and therefore also director of what was now the Yale Columbia telescope. Also present, but not in the picture, were Erica Maria Zoetelief Tromp (1902–1989) and a daughter of hers. She was the wife of the Dutch ambassador in Pretoria, where the executive branch of the South-African government was based. She was sister to the widow of Aernout de Sitter, Henriëtte

Zoetelief Tromp. The ambassador was Jan van den Berg (1899–1982), with whom Oort was apparently good friends. He had grown up in Hazerswoude and Oort maybe knew him from high school (they were about the same age, but van den Berg probably attended grammar school). After having studied Chinese language and literature, van den Berg had held various diplomatic posts in China and was now ambassador in South-Africa. Oort stayed part of the time with Jan and Erica van den Berg.

Oort discussed the future of the observatories in Johannesburg with Adriaan Wesseling, Willem van den Bos and Jan Schilt. Through Jan van den Berg he made contacts with the South-African authorities on a local and national level—he also traveled up and down to Cape Town. Nothing was decided at that stage, but it was the beginning of the move of Leiden station to Hartebeespoortdam near Pretoria, together with the Unie Sterrewag. As already mentioned, the Yale-Columbia telescope finally moved to Australia.

After that Oort had visited the Kenya expedition. First with a small plane to Vereeniging, a town south of Johannesburg, where a large lake is located. From there a large seaplane (there were few airports) from British Overseas Airways could take off, which took him to the Zambesi River near the Victoria Falls. From there to Kampalo on the north side of Lake Victoria. After two days of waiting he could take a train to Timboroa and the Kenya expedition, where he arrived at four o'clock in the morning. Van Herk had warned him that the stop was short, so he had to get up on time. Apparently he succeeded.

After this visit Oort traveled via Nairobi by plane to Khartoum and then to Cairo. He was a guest there for a few days at the Helwan Observatory, with which he, together with the director Mohammed Reda Madwar (1893–1973), made plans to work on the experiment of the zodiacal light, which had due to lack of time not been carried out in Kenya. On February 24 (1949) he was home again. This eventually resulted in a joint project in which an instrument was built in Leiden to observe the inner parts of the zodiacal light from Khartoum during the solar eclipse of February 25, 1952. Oort visited Cairo around the turn of the year (1950/51) for four weeks (probably against his wishes, but he felt he could not refuse), where he also lectured (see Fig. 8.12). The instrument was built on time, but scientific results were never published.

It was not until 1957 that Van Herk was able to publish the results of the second Kenya expedition. He received help from the Mathematical Center in Amsterdam, an institute funded by ZWO. The analysis ultimately resulted in eleven numbers, the corrections in eleven declination zones for the declinations therein in the FK3, ranging from -0.15 to +0.20 arcseconds. The astrometric satellite Hipparcos of the European Space Agency and in more recent times GAIA, have brought this work to a much higher level of accuracy. Compari-

Fig. 8.12 Oort during his visit to the Helwan Observatory near Cairo in January 1951. The other persons have not been identified, but the one on whose shoulder Oort's hand rests is probably his host Mohammed Reda Madwar. From the Oort-archives

son with Hipparcos (see [75]) indicates that van Herk's values were significant improvements. But the international astrometry community largely ignored them, probably because conservatism played an important role and these astronomers were unwilling to accept the unusual approach. I have not been able to find any references to van Herk's publication; two recent authoritative overview articles on the history of astrometry do not even mention the work! This is very disappointing considering the heroic nature of the expeditions and the enormous commitment of those involved. Perhaps Oort could have drawn more attention to it at international meetings. Van Herk was a modest man, but undoubtedly deserved more praise than he received.

8.6   The Oort Cloud

It is remarkable that Oort has had so few PhD students. He was probably too busy with his own research, directorate and national and international activities and functions. Of the fifteen, seven defended their dissertation after his formal retirement in 1970! The first three he in fact 'inherited' from de Sitter, Hertzsprung and Woltjer. Gijs van Herk was actually a student of de Sitter. His second PhD student was Leendert Binnendijk (1913–1984), who during the war worked on a dissertation under Hertzsprung. It concerned a study of the Pleiades, a nearby cluster of stars of which the brightest six or seven can be seen with the naked eye. Oort took over after the war and Binnendijk obtained his PhD in 1947. After this he worked in the United States in astrometry with Peter van de Kamp of the Sproul Observatory at Swarthmore College in Pennsylvania, and later on binary stars at the University of Pennsylvania in Philadelphia.

The third PhD student was Adriaan van Woerkom, with whom Oort had worked in the first years of the war on the vertical force in our Galaxy. Van Woerkom studied the origin of comets, a following a suggestion of Hertzsprung. Jan Woltjer was his primary supervisor (this is clear from his introduction to the thesis) and after his death Oort took over.

Comets are small bodies consisting mainly of rock, dust and frozen gases, which move in elongated orbits through the Solar System, spending most of their time far away from the Sun. The best known is Halley's Comet, which has an orbital period of about 76 years (due to disturbances of the orbit by the large planets it is slightly variable). This comet has been closest to the Sun most recently in 1986. Edmond Halley (1656–1742) had noticed that the 1682 comet had about the same orbital parameters as those of 1531 and 1608, and predicted that that comet would return in 1758. This came true, but by then Halley had already died. When comets come close to the Sun, they evaporate partially and the released gases take the dust with them, which reflects the sunlight. We then see long tails, which more or less, but not precisely point away from the Sun because of the radiation pressure of the sunlight. Charged particles are also released (induced by the Sun's ultraviolet radiation) these interact with those of the solar wind, giving another tail that points away from the Sun more accurately. After a few passages the comets usually evaporate, although larger ones like Halley can take a long time.

There must be a constant replenishment of new comets, which come from very great distances and then can end up in orbits with shorter major axes because of the gravitational pull of the large planets, especially Jupiter. An example of a comet that first entered the inner parts of the Solar System in

Fig. 8.13 Comet Kohoutek photographed by myself on 21 January 1974 with the 48 in. Schmidt telescope of Palomar Observatory. The stars are short dashes, because during the exposure the telescope followed the comet and not the stars. The tail has a length of about 6° in the sky. From the author's collection

1974 is that of Kohoutek (Fig. 8.13). Such new comets can be very spectacular, but that is difficult to predict (like Oort at the time warned, Kohoutek was indeed disappointing). By the way, Kohoutek was forced by Jupiter into an orbit such that after the passage of 1974 it left the Solar System forever. This comet has only been seen once and will never return. Van Woerkom's dissertation was to a large extent a study of the statistics of the changes of the major axis of the elliptical orbit of a comet, which occur when they pass through the inner parts of the Solar System.

If comets came from outside the Solar System, those original orbits would have to be hyperbolic rather than elliptical. If you stretch an ellipse, it becomes more and more elongated and eventually, when the orbit stretches to infinity, it becomes a parabola. On such an orbit the comet can then just reach infinity. If you stretch even more, then the comet, even if it would have come infinitely far from the Sun, would still have a velocity away from us and that is called hyperbolic. Van Woerkom had already made a careful inventory of the original orbital properties of comets seen in the inner parts of the Solar System that have done so for the first time. This is straightforward (but tedious) to determine if you know the current orbit and calculate backwards in time how it has been changed by Jupiter and the other large planets. He already concluded that there was not a single comet that seemed to come from 'beyond infinitely'.

The statistics of the original orbits of new comets indicated that they all came from distances between 20 and 150 thousand astronomical units (AE, the average distance from the Earth to the Sun). For comparison, the nearest star is 270 thousand AU. Not a single new comet was known, whose orbit was definitely hyperbolic. This made Oort postulate that there had to be a reservoir of comets in a sphere between 20,000 and 150,000 AU from the Sun. Others soon referred to it as the Oort Cloud.

There were, however, two questions to be answered. The first was, how some comets could leave that reservoir and move to the inner parts of the Solar System. Oort showed it was reasonable to assume that disturbances caused by the gravitational force of nearby stars could be responsible for this. But the most difficult question was how that cloud was created in the first place. For that, Oort had the idea, that they would have come out of the asteroid belt between Mars and Jupiter. Asteroids are fragments of a 'failed' planet with dimensions up to several tens of km (and a small number of even larger ones). Together they represent a mass of less than a thousandth of the Earth and perhaps a hundredth of Mars. So there must have been many more in the early stages of the Solar System, of which Oort then speculated that some of them now form the cloud of comets. He showed that Jupiter, and to a lesser extent Saturn, could have pulled them into elongated orbits, usually in a few steps, where passing stars would then have pulled them into more circular orbits. This would also explain why comets come from all possible directions and are not limited to close to the plane of the planets. He also figured out what the distribution of the major axes of orbits of new comets would then have to be. That was quite correct, except that there would be too few 'new' comets compared to 'old' comets. An inventory by student Maarten Schmidt gave the solution; he found that when comets have come close to the Sun once (or maybe a few times), the next time they will brighten much more quickly and evaporation will be faster. Therefore, new comets like Kohoutek are often less spectacular than predicted. The theory of the Oort Cloud was quickly accepted as convincing and remains so to this day. In 1951 Oort was asked to give the Halley Lecture at Oxford University for the second time. Oort decided to talk about the origin of comets and so his Halley lecture of May 1, 1951, was the first one that dealt with comets.

8.7 The Oort Family

After the war, the children of the Oorts picked up their education. The eldest son Coen went to Leiden to study economics after secondary school. This study was a specialization within the Faculty of Law. He continued to live at the Sterrewacht on the top floor of the director's house. Daughter Marijke

Fig. 8.14 Te Oorts with two of their children, Marijke and Abraham Hans (Bram) in 1956. The other son Coenraad Jan (Coen) lived and studied in the United States at the time. From the Oort-archives

went to the conservatory in Den Haag and the youngest son Bram went to study physics in Leiden in 1952 (Fig. 8.14).

Oort's youngest son Bram has made some documents available to me, from which I quote parts. I start with Bram Oort's notes about his brother and sister.

'I am going back to my memories of my brother Coen. After obtaining his degree in economics, he went to Chicago in 1962 to continue his education there. He was probably influenced by Milton Friedman in Chicago and belonged to the Chicago school – that promoted free trade and less government. In Chicago he met a younger student, Marianne Lissy, whom he later married. They came to the Netherlands together and lived on the top floor of my parents' house at the Sterrewacht. Two sons, Marc and Maarten, were born there. It was a great joy for my mother to be able to take care of her grandchildren. My brother obtained a PhD in economics and got a job as a professor at the University of Utrecht. Later he became Treasurer-General of the Netherlands (1971–1977), and ultimately a top banker of the ABN-AMRO Bank. As a banker he earned an excellent salary. For him money was never a problem.

My sister Marijke worshiped her three-year older brother. They were very fond of each other and I often felt excluded as the youngest brother. In high school, Marijke loved languages and was captivated by singing and piano. She studied singing and piano at the Conservatory of Music in Den Haag. She taught amateur singers and gave piano lessons. Later, to her great regret, she had problems with her hearing until eventually she could not hear much anymore. Hearing problems are very common in the Oort family; my father, Coen, Marijke and I have been severely handicapped by them. Fortunately I profit from recent developments in the field of hearing aids. Eventually my sister broadened her music studies and started to play the harp.'

The following quotation is part of the speech that daughter Marijke gave at the funeral of her brother Coen in 2006:

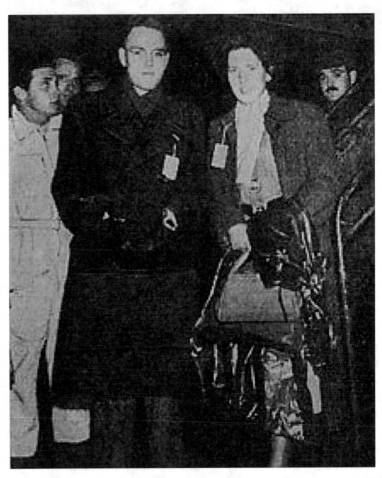

Fig. 8.15 Oort's oldest son Coenraad (Coen) en daughter Marijke upon arrival in Rome in December 1950 after a KLM DC6 flight. Coen had won a trip for two in an essay contest and had chosen to take his sister with him. It is not mentioned from which newspaper this clipping came. From the Oort Archives

'When we returned to Leiden in '45, to our Sterrewachthuis, we all had quite some adjustment problems.

Coen went to university, lived on the top floor of the house. Every half hour he came thundering down all the stairs to drink a cop of tea or do something else. My mother and I always wondered if he was actually studying, because all the time he appeared again. But apparently a short moment of study was enough for him.

When he received a prize from the radio for an essay on Europe at that time, he won a trip to Rome for two people. He left us until a week before departure in limbo as to who would go with him. Only then did it become clear that I was going with him and not some unknown, fantasized girlfriend. We had a lot of fun together, were like two very strange ducks at a reception at the embassy and at an audience with the Pope.

From the States, where Coen studied for a year, we learned on the very last page of a long letter, that he had a girlfriend he was going to marry and that he was going to take her with him to the Netherlands. Everybody advised Coen to take Marianne on a trial run (maybe she wouldn't like to live in Holland). Despite all the good advice, Marianne and Coen married in America in the same week as we did: they in America and we in Leiden. They also went to live again on the top floor of the Sterrewacht house. Marianne enjoyed life in Leiden and conquered everyone's heart in no time.

We had a wonderful time together with the four of us, sat together almost every weekend. Marianne already cooked the most delicious meals and we did all kinds of games, supposedly to teach Marianne Dutch.'

The essay competition that Coen had won was the ERP Prize (ERP is European Recovery Program, better known as the Marshall Aid). It was organized by the KRO—a radio broadcasting organization—and Coen had to choose between London, Paris or Rome. The winner had already been announced in the spring and Coen took six months to reveal who was allowed to come with him (see Fig. 8.15).

At the funeral of Marijke Oort in 2008, Bram Oort said among other things:

'As a little girl, Marijke was very sporty; loved water, dived from the high board, and swam like an otter. Later on we sailed in our little cabin boat the 'Bounty'. I remember a sleepless thunderstorm-night with Marijke on the side of Lake Braassemer; buzzing mosquitoes, pouring rain and lightning. Another adventure was the sailing trip to the river IJ in Amsterdam where we spent the night in the harbor near the red light district. The next day we sailed with Father via Enkhuizen across the IJsselmeer – quite a daring undertaking with an unseaworthy boat. We spent many summers sailing like this, often with the Nauta family in their 'Boek-Boek'.

After returning to the Observatory, you went to the HBS for girls with your love for languages and literature. That is when your love for music developed. I have good memories of how the Sterrewacht was filled by your singing and piano playing, and of your later performances at the family parties. But at this time a new phase started with Tom's entry into our family. I remember the enormous bouquets with which Tom courted Marijke. Then your long engagement, and getting married in '54. Very unexpectedly, Coen was ahead of you and married his love, Marianne, a few days earlier in Chicago.'

And Bram Oort about his mother:

'One of my Mother's characteristics was her love of poetry and art. Her father was a poet. It seems that the love for art was developed by herself. Because Father worked so much, she had time to read and study. Together they went for many years after WWII to an art course given by H. Bremmer (a painter who drew the Kröller–Müllers' attention to the work of van Gogh; Bremmer was one of the discoverers of van Gogh). Many of Van Gogh's paintings can now be seen in the Kröller–Müller Museum in the Hoge Veluwe. As an anecdote: my grandmother Oort knew the van Gogh family; she was offered the opportunity to be given some of his drawings. But she did not like them well enough and did not accept the offer – this to the chagrin of her descendants!

Bremmer also advised my parents to buy paintings and sculptures they liked. In our house we have two bronze statues that belonged to my parents; a statue of a deer running away – made by Radeker. It is a first study for a life-size statue in stone that is now standing in the middle of the Kröller–Müller. My parents loved it; it stood in the middle of the coffee table. Now it is in the same place with us. [...]

My mother was, in a way, the 'soul' of the Sterrewacht. The astronomers liked to come and have tea or dinner with us. My mother was always ready to serve tea, biscuits, etcetera. Many friends loved her because of her cheerful nature and cordiality. Father learned so much from her; without her he might have become a man of science in an ivory tower. Together they gave a lot of parties for the astronomers and students of the Sterrewacht – I remember guests playing a live chess game a few times on the big tiles in the middle of our house (Sterrewacht 5).'

Vincent van Gogh does not need introduction. The art collector Helene Emma Laura Juliane Müller (1869–1939), who founded the Kröller–Müller Museum with her husband, businessman Anthony George (Anton) Kröller (1862–1941), was advised by Hendricus Petrus (Henk) Bremmer (1871–1956). Bremmer was a painter himself but better known as an art critic, educator, collector and art dealer. The sculptor is Johannes Anton (John) Rädecker (1885–1956); for some of his sculptures, including the aforementioned Hinde, see [97]. Jumping ahead to a later chapter: Fig. 12.5 shows the Oorts in the 1980s with this sculpture. It is perhaps worth mentioning that de Oorts read

Fig. 8.16 Coen Oort on his way to the defense of his dissertation in 1958 in Leiden. Next to him are his wife Marianne and his eldest son Marc. Photo made available by Marc Oort

to each other in the evening the works of William Shakespeare (1564–1616) (Fig. 8.16).

Here is a brief description of the lives of the children Oort. Coen Oort went to Chicago in 1954 on a Fulbright Fellowship, where he obtained a masters degree. There—as already mentioned—he met his future wife Marianne Stazie Lissy (b. 1935). They married in 1954. After his return from Chicago Coen worked on his dissertation while they lived on the upper floor of his parents' house. The first two sons were born there, including the eldest, Marc Jan Anton (b. 1958), who was eventually to become an astronomer. In 1958, Coen defended his dissertation *Decreasing costs as a problem of welfare economics*, and subsequently worked at the Ministry of Economic Affairs in Den Haag. In

1960, he was appointed professor of economics and statistics at the Faculty of Law of Utrecht University, where he remained until 1971. He was then appointed Treasurer-General at the Ministry of Finance, which lasted until 1977. He held various positions at ABN-AMRO Bank. His wife had studied art history at the University of Chicago. Later in Leiden she studied English language and literature (bachelor), Sanskrit and the history and culture of South Asia (PhD 1980). She worked as a librarian at the Department of History and Culture of South Asia.

Marijke Oort gave music lessons after her studies at the conservatory. Later she also started to play the harp. In 1954 she married Jacobus Thomas (Tom) de Smidt (1923–2013). Tom de Smidt had studied law and had a special interest in the history of Dutch jurisprudence. After his graduation in 1954, he worked for the Dutch parliament until he was appointed professor in 1956 of 'introduction of jurisprudence and the history of Dutch law' at the University of Amsterdam. From 1965 he combined this with an appointment in Leiden on the same subjects.

Bram Oort studied physics in Leiden. After his PhD and military service he moved to the Massachusetts Institute of Technology MIT to study meteorology and oceanography. In 1964 he returned to the Netherlands where he obtained his PhD thesis *On the energy cycle in the lower stratosphere*, in Utrecht. He married Bineke Pel (b. 1937), who had studied history. Bram worked at the KNMI (the Royal Netherlands Meteorological Institute) until in 1966 they emigrated to the United States, where Bram worked at the Geophysical Fluid Dynamics Laboratory, first in Washington, D.C., but after two years in Princeton. He was a visiting professor at Princeton University, and published on the observed energy and water cycles in the atmosphere. He summarized his work in a book on the *Physics of climate*, written with Jose P. Peixoto (1922–1996) of the University of Lisbon. His wife taught at the Art Museum at Princeton University, studied social work at Rutgers University in Newark, New Jersey, and practiced synergy therapy. After Bram's retirement in 1996, they moved to Vermont, where Bram taught macrobiotic shiatsu at the Kushi Institute.

Fig. 9.0 Oort in 1948. From the Oort Archives

9

From Kootwijk to Dwingeloo

We should do astronomy because it is beautiful and because it is fun.
We should do it because people want to know.
We want to know our place in the Universe and how things happen.

John Norris Bahcall (1934–2005)

Look at the stars. It won't fix the economy. It won't stop wars.
It won't give you flat abs, or better sex or even help you figure out
your relationship and what you want to do with your life.
But it's important. It helps you remember that you and your problems
are both infinitesimally small and conversely,
that you are a piece of an amazing and vast Universe.

Kate Bartolotta (1976–2018)

9.1   The Discovery of the 21-cm Line

Henk van de Hulst's prediction that the spectral line of neutral hydrogen at 21 cm wavelengths would be strong enough to detect, created a unique opportunity to study the structure of our Galaxy. A few months after the war, Oort sent a letter to Grote Reber, via Otto Struve of Yerkes Observatory (closest to where Reber lived), asking what his telescope had cost. Reber wrote back and listed in detail his expenses, which amounted to several thousand dollars. Oort contacted Bakker of Philips NatLab (he had also spoken at the colloquium in

---

John Bahcall was an American physicist and astrophysicist.
Kate Bartolotta, American author and self-care coach.

© The Editor(s) (if applicable) and The Author(s), under exclusive license
to Springer Nature Switzerland AG 2021
P. C. van der Kruit, *Master of Galactic Astronomy: A Biography of Jan Hendrik Oort*,
Springer Biographies, https://doi.org/10.1007/978-3-030-55548-1_9

1944), who had contacts at machine factory and railway builder 'Werkspoor', where a first design was quickly made for a radio telescope with a diameter of 25 m.

But where was the money to be found to build this? The first Prime Minister after the war (from June 1945 to July 1946 to be precise) was Willem Schermerhorn (1894–1977), professor of civil engineering and geodesy at the Technical University Delft. Oort apparently knew him personally, because in the correspondence that followed, they used first names. It resulted in a visit by Oort to Schermerhorn on November 16, 1945. In doing so Oort ignored the appropriate channels and in particular passed curators of the University of Leiden and the relevant minister. It was not uncommon to do business directly with 'Den Haag' (ministers and other higher officials), although the higher administrators of universities opposed this and officials at the ministries increasingly tried to prevent it. The Prime Minister did certainly like Oort's ideas for a radio telescope and the result was that Oort would write a proposal and have it assessed by the Royal Academy of Sciences (KNAW). In due course the proposal ended up with the Minister of Education, Arts and Sciences, Gerardus van der Leeuw (1890–1950), a professor of theology from Groningen. The application was for 100,000 guilders (in purchasing power 490,000€ today), but such amounts of money were of course not immediately available for pure science under the circumstances.

One of the directors of Philips NatLab, Herre Rinia (1905–1985), had pointed out to Marcel Minnaert in Utrecht that German defense radar antenna's, part of the Atlantic Wall, had been left behind in the dunes. These Würzburg–Riesen had a diameter of 7.5 m. The head of the PTT's[1] department for radio service, Anthonet Hugo de Voogt (1892–1969), was interested in long-range radio transmission and the effects of the ionosphere—a higher layer in the atmosphere where charged particles interfere with and reflect radio waves – and its relationship with solar activity. Marcel Minnaert and his Utrecht deputy-director Jacob (Jaap) Houtgast (1908–1982), were also interested in this. In this way de Voogt came into contact with Oort and through de Voogt the PTT made a Würzburg–Riese available for Dutch astronomy at the PTT site in Kootwijk near Apeldoorn. This was of course much more realistic than immediately building a large, expensive radio-telescope while the 21 cm line had not even been discovered (Figs. 9.1 and 9.2).

The appropriate route of obtaining financing of science was the aforementioned Netherlands Organization for Pure Scientific Research ZWO, which had been established in 1947 along lines drawn up by Schermerhorn. Because

---

[1] Post, Telegraph and Telephone, the national organization that was responsible for providing channels for communications.

Fig. 9.1 Photograph from 1951 of the Würzburg–Riese 7.5 m antenna, which was placed in Kootwijk to realize the first radio observatory in the Netherlands. With thanks to CAMRAS/ASTRON, C.A. Muller Radio Astronomisch Station, which uses the Dwingeloo Telescope for amateur purposes (such as 'Moon-bouncing') and to Ard Hartsuijker. [98]

Reber had built his own radio-telescope in his backyard, Oort assumed that the workshop of the Sterrewacht should be able to do the same. The insight soon came that this was unrealistic and in 1949 this resulted in the establishment of a special foundation for radio astronomy, which was called the 'Stichting Radiostraling van Zon en Melkweg' RZM (Foundation for Radio-radiation from Sun and Milky Way, but in English referred to as Netherlands Foundation for Radio Astronomy, NFRA). This turned out to be of vital importance because it brought together several important players that were necessary for a successful approach. This was evident from the composition of the RZM board. Oort was chairman and astronomical members were Henk van de Hulst (lecturer in Leiden since 1948), and Marcel Minnaert and Jaap Houtgast from Utrecht. Philips Natlab was represented by Herre Rinia and Frans Louis Henri Marie Stumpers (1911–2003), a radio engineer. De Voogt was a member for the PTT and there were two persons from the KNMI (the Royal Netherlands Meteorological Institute, both geophysicists: Jan Veldkamp (1909–1994), who worked on ionospheric issues and (paleo)magnetism, and Felix Andries Vening Meinesz (1887–1966), director and specialist in gravity measurements.

In 1949, RZM applied to ZWO for 200,000 guilders for the construction of a 25 m radio-telescope. But this was not awarded; understandably because it was 16% of the ZWO annual budget (of 1.25 million guilders) and the line

Fig. 9.2 Lex Muller on the stairs of the Kootwijk radio-telescope. From the Archives of the Sterrewacht Leiden

had still not even been detected. But with the Würzburg antenna the latter could be remedied. Leiden physicist Cornelis Jacobus Gorter (1907–1980) had a modified radar receiver from excess American stock and an engineer H. Hoo would install it for observation of the 21-cm line. But unfortunately, in March 1950 a fire destroyed the receiver.

Hoo was replaced by another engineer, Christiaan Alexander (Lex) Muller (1923–2004). Stumpers (from the RZM board) took care of the parts for a new receiver with the help of Philips NatLab. Meanwhile Henk van de Hulst spent some time at Harvard University to give a course in radio astronomy (organized by Bart Bok, who also recognized the potential of radio observations). This was a fortunate circumstance, because Edward Mills Purcell (1912–1997) worked at Harvard and he was also looking for the 21-cm line. He was an electrical engineer, one of the co-discoverers of the NMR (nuclear magnetic resonance), on which MRI (magnetic resonance imaging, without the nuclear of NMR, because it reminds of nuclear energy and radioactivity) is based. Purcell shared

Fig. 9.3 The receiver for the Kootwijk radio-telescope with which the 21-cm line was co-discovered. The person sitting in the chair is Lex Muller. From the Oort Archives

the Nobel prize for physics in 1952. He had read in the not very accessible Russian publications of Iosif Samuilovich Shklovsky (1916–1985) about his independent prediction of the line. A student of Purcell, Harold Irving Ewen (1922–2015), was commissioned to search for that line as a thesis subject. This became a fine example of collaboration in science; the communication between the Dutch and the American group was open and there was no real race to be the first.

Van de Hulst played a mediating role. It was clear to Lex Muller that receivers were too unstable for direct measurement, but had to be compared continuously (several times per second) to a stable noise source in the telescope. This process came from Robert Henry Dicke (1916–1997), who had applied this principle during the war in research into the stability of radar techniques. It is called a Dicke switch and Muller had built such a Dicke-switch receiver (see Fig. 9.3). But it still was not stable enough. Now Purcell and Ewen had figured out that it would be better not to compare the signal from the sky with that from a local noise source, but with the sky itself at a nearby frequency. Atmospheric effects and the like would then be comparable in both signals and cancel. They told this to Henk van de Hulst. Conversely the latter realized, that

Purcell and Ewen applied far too small a switch in frequency for the comparison signal. Along a line of sight through the Milky Way we see hydrogen at a large range in velocities relative to us, so the 21-cm line radiation is distributed in frequency over a certain area. Purcell and Ewen knew nothing about astronomy and thought of a very narrow line as in laboratory experiments. Van de Hulst on the one hand advised them to take a bigger step in frequency and on the other he told Muller of the trick to do the Dicke switch in frequency. Ewen and Purcell discovered the 21-cm line on March 25, 1951. Muller needed time to modify his receiver for a frequency switch and found the line on May 11, 1951, less than seven weeks later. The results were published at the same time on the same page in *Nature* magazine, together with a telegram from Australia, confirming it once again.

The article by Muller and Oort contained right away the first astronomical analysis. Oort interpreted the measurement as a determination of the rotation velocity of the Milky Way System halfway between us and the center as being about 190 km/sec. But further analysis had to wait until Muller had built a more sensitive and stable receiver, and that took, for Oort frustratingly long, until June 1952.

9.2   Spiral Structure of the Galaxy

At Kootwijk the pointing of the telescope was done turning cranks by hand; the results were recorded on a strip-chart recorder and measured up and reduced in Leiden. The work was mainly done by students, such as Maarten Schmidt and Gart Westerhout (1927–2012) (see Fig. 9.4). The analysis made use of the fact that hydrogen will have a different radial velocity (away from us or towards us) at different positions along the line of sight. This is explained in Fig. 9.5. The Sun is at point S and the center of the Galaxy at GC. Here the line of sight has a direction corresponding to a Galactic longitude l (the angle at the top at S). The rotation is taken the same everywhere (in km/sec) and that is a good approximation. The rotation projects onto the line of sight with a varying magnitude. Seen from us, the radial velocity along this line of sight is everywhere that projected velocity minus the projection of ours at S. We then see that when we go to A a certain velocity remains which is away from us; at N it is maximal and then decreases again. At B it is the same again as in A, so we cannot distinguish hydrogen from those two points from each other. At S' the difference is zero and at C the velocity is towards us. So if

Fig. 9.4 Maarten Schmidt (left) and Gart Westerhout at the Kootwijk radio-telescope. With thanks to Ard Hartsuijker; from the Archives of ASTRON Dwingeloo

in the observation at this longitude we see a peak in the 21-cm line-radiation (which then was supposed to be a spiral arm) one can find out with the velocity where that hydrogen is along that line. In the subcentral point N the rotation is entirely in the line of sight, and by finding the maximum measured velocity at various longitudes, one can determine the rotation velocity as a function of distance from the center.

For gas further out than the Sun, you need a model for the distribution of mass in the Milky Way to calculate what the rotation velocity is. Such a model was developed by Maarten Schmidt. He collected information about the distribution of various components and brought it together in four ellipsoids with density distributions, in such a way that all constraints of known factors (rotation velocity at the Sun, distance from the center, Oort constants, etc.) were met. This gave, among other things, the rotation velocity at distances from the center greater than that of the Sun. It formed his dissertation, on which he obtained his doctorate in 1956 with Oort as supervisor.

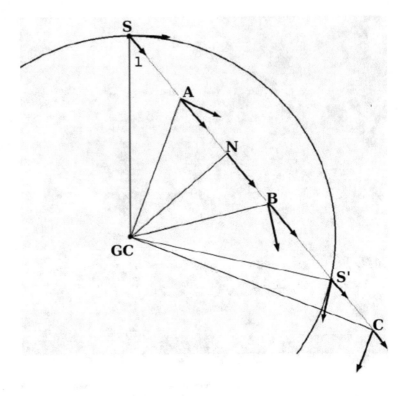

Fig. 9.5 Geometry of HI (neutral hydrogen) observations. Gas at different positions along the indicated line of sight has different projected velocities and therefore different radial velocities relative to the Sun. Figure by the author

Eventually, this method was used to create a map of our Galaxy. This is shown in Fig. 9.6 first published in 1957, based on work—besides that of the authors Muller, van de Hulst en Oort—by various students, in particular Maarten Schmidt, Gart Westerhout and Kwee Kiem King (b. 1927). There was no doubt that our Galaxy had spiral structure (Fig. 9.7).

The discovery of spiral structure in the Galaxy was a monumental result. The website of the Nobel Foundation makes public the nominations until (at the time of writing) 1966 [100]. Oort, and Muller were nominated for the 1956 Nobel Prize in Physics by Lindblad, with support from the Swedish astronomer and solar physicist Karl Yngve Öhman (1903–1988) and of the Delft physicist Hendrik Berend Dorgelo (1894–1961). Lindblad and Öhman repeated this in 1957, 1958 and 1961. Astronomy at that time, however, was too far removed from physics to be considered. Hubble has also been nominated for the discovery of the expansion of the Universe, but he also did not receive the prize. Oort was also nominated by himself, but that was not

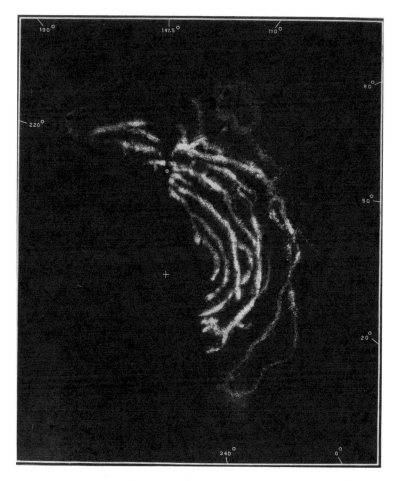

Fig. 9.6 Distribution of neutral hydrogen in the Galaxy from the Kootwijk survey. The density in hydrogen in the spiral arms is 1–2 hydrogen atoms per cm^3 or more. The cones in the center-anticenter directions are empty; all hydrogen there moves perpendicular to the line of sight so that distances cannot be determined. The ambiguity A versus B in Fig. 9.5 is solved by dividing the gas evenly over both positions. The left part is empty, because that part of the Milky Way is only visible from the southern hemisphere. From the Oort Archives, published in BAN 475 (in 1957) [99]

successful either: three times (1958, 1959 and 1960) by Otto Struve, in 1956 by Abraham Cornelis Sebastiaan van Heel (1899–1966), professor of optics in Delft, and again in 1960 by van Heel together with seven other physicists from Delft. Brute force did not help either.

Fig. 9.7 Oort and van de Hulst at the time of the first measurements of the 21-cm line in the Milky Way with the Kootwijk radio telescope. The blackboard shows a sketch of the method and the first results. From the Archives of the Sterrewacht Leiden

9.3   Pasadena, Princeton and Gas Clouds

In 1950, Oort began making plans to go to the United States again for a longer period of time. Of course Pasadena and the observatories of Mount Wilson and now Palomar Mountain attracted him. In May 1950 Oort made contact with Walter Baade, with whom he had become good friends. The Rockefeller Foundation which had financed the 200 in. Telescope, had donated it to the California Institute of Technology in Pasadena. This university, often referred to as Caltech, had grown into a research and educational institution of stature. The observatories of Mount Wilson and Palomar Mountain were administratively merged into a single organization, the Mount Wilson and Palomar Observatories, whose operations were jointly funded by Caltech and the Carnegie Institution of Washington (which according to some, by symmetry should be referred to as Carwash). To this end, Caltech established in 1949 an astronomy department to carry out research and provide teaching in astronomy. As director of this department they hired Jesse Greenstein. Greenstein, whom we already met in connection with the properties of absorbing dust in interstellar space, had been working at the Mount Wilson Observatory since 1948.

Greenstein wanted Oort to spent some time in Pasadena to teach at Caltech. But there was also interest expressed from Princeton to have Oort visit there for some time. Russell had been succeeded there as director by Lyman Strong Spitzer (1914–1997) and on the staff was also appointed Martin Schwarzschild (1912–1997), the son of Karl Schwarzschild. His father was Jewish and this was reason to flee Nazi Germany in the 1930s. Spitzer and Schwarzschild together ran the astronomy department in Princeton and formed a duo of very influential astronomers, each also having an enormous reputation individually. Oort had met Spitzer at a conference in Paris organized by Henk van de Hulst on the interstellar medium, after which they kept up regular correspondence on the subject.

Greenstein and Spitzer coordinated an offer to Oort to come to both places for some time to lecture and to set up joint research projects. They wrote Oort on two consecutive days in October 1950. In the course of 1951 the plans became more concrete. Mrs. Oort would accompany her husband again this time. When others got wind of the plans, the opportunity was seized by the American Astronomical Society to invite Oort to give the prestigious Henry Norris Russell Lecture. This would take place during the winter meeting of the AAS, which was held late December 1951 in Cleveland, Ohio.

The Oorts left for America before Christmas 1951 and after a rough crossing, which had made it difficult for Oort to prepare his Russell Lecture, they stayed with Jan and Jo Schilt for a few days. From there by train to Chicago, where their son Coen was studying. Despite the hours of delay the train had, he was still waiting for them at the railway station. From there to Cleveland, where Oort gave his Russell Lecture on December 27th. On January 1, 1952 they traveled by plane to Los Angeles. Flying at that time was still unusual; they were delayed by 'fog, busy traffic and headwinds', so they had to make an extra stopover in Las Vegas. In Pasadena they visited many colleagues and at the end of January they stayed for a few days on top of Mount Wilson in the Kapteyn Cottage. Oort had an observing run on the 100 in. Telescope together with staff member Rudolph Minkowski (1895–1976), an astronomer from Germany, who was a Christian, but of Jewish descent, and yet had to flee the Hitler regime. He had worked in Hamburg with Baade, who had arranged for him to work on Mount Wilson. Minkowski developed into a renowned spectroscopist. What Oort and he observed is not clear; it seems obvious that they tried to obtain a better determination of the rotation curve of NGC 3115, because its dynamics were still an important point of attention for Oort. If so, it did not lead to any important new information. There is mention in the Archives of a second observation run during Oort's stay in Pasadena.

The Oorts had an excellent stay in California. They visited Death Valley and they made an extensive trip to San Francisco and the surrounding area. Otto Struve, who was now in Berkeley, picked them up in Pasadena in the official observatory car (Oort wrote in his diary that he drove very fast, as much as 80 miles (about 125 km/h) and they stayed with Donald Shane and his wife Mary Lea Heger (1897–1983), also an astronomer, on Mount Hamilton, the site of Lick Observatory. Upon their return to Pasadena they began what Oort called 'our grand journey to the East' in his diary. They traveled by car; a Ford, which they apparently had bought for the trip. They left on March 23rd and in Arizona they teamed up with their son Coen, who had traveled by train from Chicago. They visited Grand Canyon, in which they descended on foot as far as the Colorado River. By car they drove on to Williams Bay, Wisconsin (Yerkes Observatory of the University of Chicago), where they stayed for a few days with their good friends Bengt Strömgren and his wife Sigrid. The journey ended for the time being in Princeton on April 11th. Oort had kept all details about the car trip in his diary; they had driven 4537 miles (7301 km) since leaving Pasadena. They had only had a flat tire twice in Arizona, had to have lights replaced once and had to have a repair done for a leaking water hose. Oort recorded a total fuel consumption of 308 gallons (1167 liters). This is a to modern standards a very expensive use of only 6.25 km per liter.

Oort was fascinated by the question of how the clouds of interstellar gas maintained their velocities. After all, regular collisions between gas clouds, which are 'inelastic' (i.e. not like billiard balls), would reduce the mutual motion, heat the gas and radiate away energy. Oort focused on areas of star formation and the energy emitted by the young hot stars. Spherical regions of ionized gas, described by Bengt Strömgren, form around such stars. An example of such a Strömgren sphere can be seen in the Rosetta Nebula (see Fig. 9.8) with at its center the cluster NGC 2244, which contains many hot O-stars. The luminous area is the Strömgren sphere (the central hole is filled with very hot gas as a result of the star winds—outflow of hot gas—from the central bright stars). Oort and Spitzer now looked at the situation in case the medium is not uniform (unlike what Strömgren did). An example of a situation they studied is that of a cloud of gas near an O star (or a number of such stars). This heats and ionizes one side of the cloud and the gas there

Fig. 9.8 The Rosetta Nebula and the central cluster NGC 2244. The colors are 'false' and correspond to light from different atoms or ions. Light from neutral hydrogen is coded red, from doubly ionized oxygen green and from singly ionized sulfur blue. Credit: National Optical Astronomy Observatory/Association of Universities for Research in Astronomy/National Science Foundation [101]

will expand. This gives a push to the gas cloud as a kind of rocket propulsion. They looked at several more situations and came to the conclusion that this 'Oort-Spitzer mechanism' can invoke considerable increases in the velocities of gas clouds in the interstellar medium.

In a separate article by only himself, Oort estimated the amount of energy lost and radiated in the interstellar medium in collisions and how much is added by the Oort-Spitzer mechanism. He found that the two are roughly equal. He also described how expansions in such areas can give rise to star formation by compression at the interface between the expanding gas and the cooler interstellar medium. Those stars then have considerable velocities. This explained the high velocities that O stars sometimes have (up to 50 or 70 km/sec) in the so-called OB-associations, expanding star clusters with many young O and B stars.

This work was typical of Oort. He made schematic models and rough estimates and convinced himself that it could be a realistic explanation. But then he lost interest and left the working it all out in details to others.

9.4   Back to the Crab

The identification of the Crab Nebula as the remnant of a 1054 supernova had aroused Oort's interest in this object. Now, it turned out to be a source of radio radiation as well. In the early days of radio astronomy, it was soon found that there were discrete sources of radio radiation. These were indicated by the name of the constellation in which they occurred, the brightest followed by a letter 'A'. The Galactic Center was an example, which called Sagittarius A. The Crab Nebula was Taurus A. Cassiopeia A also turned out to be a supernova remnant; nebulous filaments indicating expansion and showed that the supernova must have exploded around 1670, but due to absorption this was not observed. But the radio sources could also be extragalactic systems, like for example Virgo A, the giant elliptical galaxy, M87 (see Fig. 6.4), in the center of the Virgo cluster, a large cluster of galaxies at about 15 Mpc. Cygnus A turned out to be a very distant galaxy. We now know that this indicates activity in the nuclei of large galaxies. The question was, of course, what caused the Crab Nebula, and these other sources, to emit radio radiation.

Oort now wanted to study the expansion of the Crab Nebula by measuring how the surface brightness of the amorphous part decreased with time (see Fig. 8.4). Now there was a young astronomer, Théodore (Fjeda) Walraven (1916–2008), who had studied astronomy in Amsterdam and was an expert in building instrumentation for optical telescopes. He had worked with the Rockefeller astrograph in South-Africa and had been involved in the construction of a telescope in Leiden, the Zunderman telescope of the Sterrewacht. This is a 46-cm reflector telescope, named after the head of the workshop of the Sterrewacht, Hermanus Zunderman (1880–1969). When it was finished in 1947, Walraven had built a photometer for it, an instrument that can measure the brightness of stars with a photocell. Oort entrusted him as well with the building an instrument to measure the surface brightness in the Crab Nebula. They used the 33-cm photographic telescope, (see Fig. 9.9) where Walraven replaced the plateholder with a photometer and wheel with filters to select wavelength areas. The telescope is a double one, the smaller of the two, which has a mirror with a diameter of 15 cm, to precisely track the objects in the sky by holding a star at the intersection of two perpendicular cross-wires. The photometer contained an instrument to detect and electronically amplify the light (a so-called photomultiplier).

Before they were up and running, Oort and Walraven heard that astronomers in the Soviet Union had found that the light of the Crab Nebula was highly polarized. Light can be understood as a wave, which can be described with a periodically changing electric and magnetic field, such that when the one is at

Fig. 9.9 The Photographic Telescope of the Sterrewacht Leiden, which Oort and Walraven used to measure the polarization in the Crab Nebula. It has an objective lens with a diameter of 33 cm and a focal length of 3.4 m. The tracking viewer is in the same tube and has an aperture of 15 cm. The 'English mount' is somewhat unusual; the construction is simple and the telescope is easy to move, but the disadvantage is that areas near the pole cannot be observed. The best known telescope with this mount is the 100 in. at Mount Wilson. From the Archives of the Sterrewacht Leiden

maximum when the other is zero. A wave corresponds to a certain plane. The planes belonging to the oscillation in the electric and the one in the magnetic field are perpendicular to each other. The plane of the wave of the electric field is usually taken as that of the light. It now may be that within any small area the polarizations are in all directions; then the light is unpolarized (0%). If everything has the same direction of polarization, the polarization is 100%.

Walraven immediately replaced three of the filters that selected wavelengths with polaroid filters, which let through polarizations that differed by 60°. If a position is observed successively through all three filters, the degree and direction of polarization can be determined. Fig. 9.10 shows the results. The Crab Nebula was measured there with an aperture of half a minute of arc. A

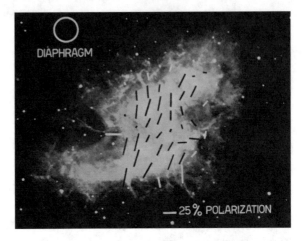

Fig. 9.10 Polarization of the light from the Crab Nebula observed with the Photographic Telescope of the Sterrewacht Leiden. Oort and Walraven used a special photometer, and the measurements were performed with an aperture of 0.5 arc min. The length of the lines indicates the degree (percentage) of polarization and the direction that of the (projected) electric field of the light. The magnetic field in the Crab Nebula is perpendicular to that. From the Oort Archives

remarkable result, all the more so when it is realized that this was obtained in the middle of Leiden with poor observing conditions and much light pollution! It shows that there is systematic polarization over areas of half a minute of arc.

What does it mean that the light is polarized and what mechanism did give rise to it? As I discussed above, the radio radiation of the Milky Way was initially conceived as free-free emission from areas with ionized gas. This was radiation emitted when in an area of space ionized hydrogen a free electron comes close to a proton without recombination. Even before the discovery of the 21-cm line, Gart Westerhout and Oort had interpreted the radiation from the Milky Way (and thus from the discrete sources) as such free-free radiation, partly because no other reasonable mechanism was known. They assumed that it had to do with stars and that therefore the distribution on the sky corresponded with that of the stars in the Milky Way Galaxy. What they did not know was that already in 1950 it had been suggested that the radio radiation could also be so-called synchrotron radiation, resulting from highly energetic electrons moving in a magnetic field. They then spiral around the magnetic field lines and the forced change of direction gives rise to observable radiation, that is to say, if those electrons are so energetic that they have velocities of the order of that of light. This hypothesis originates from the Swedish (later winner of the Nobel Prize) Hannes Olof Gösta Alfvén (1908–1995) and Axel Nicolai Herlofson (1916–2004). ⏤

In the first particle accelerators, cyclotrons, there was a magnetic field that kept the (charged) particles in circular orbits. The radius of that orbit depends on the energy or velocity of the particle (say an electron), but the orbital time does not. At a certain position in the orbit one then used an electric field to accelerate the particles. Half an orbit further that happens again, but then the electric field must be reversed in time to give an acceleration again. When larger and larger cyclotrons were built and the energy of the particles became higher (so that they moved at velocities approaching the speed of light), the period in the orbit of the particles became dependent upon the energy and the synchronization was lost. So one had to change the period of changing of the electric field to go in step with the orbital time. This was then called a synchrotron. In these devices a glow was observed (later in cyclotrons as well), which came from the electrons in curved orbits. This was called synchrotron radiation and was highly polarized.

Radio radiation from the Milky Way is largely synchrotron radiation. The interstellar space is filled with a magnetic field and very energetic electrons. Now Iosif Shklovsky, whom we already met as an independent predictor of the observability the 21-cm line, proposed that the radiation from the amorphous part of the Crab Nebula should be optical synchrotron radiation. This was the explanation of the optical polarization that Oort and Walraven had found. Oort estimated the strength of the magnetic field and the energy of the electrons. The total energy in the nebula should then be 3×10^{41} joules. Oort wondered if this was reasonable, and compared it to the energy released when the hydrogen in the Sun would be converted entirely into helium, which came to $\sim 10^{45}$ joules. He also estimated that the electrons would lose their energy in about 200 years.

The remnants of a star that has become a supernova can be a neutron star after the drastic contraction. Because a large part of the angular momentum—the momentum of rotation—is conserved, it rotates extremely fast; periods can decrease to significantly less than a second. It will also develop a strong magnetic field and therefore will emit electromagnetic radiation in bundles along the magnetic poles. The resulting 'lighthouse object' can then be seen as rapidly successive pulses, at least if we happen to be in that beam. These are pulsars, which were discovered in 1967 by Susan Jocelyn Bell (b. 1943) and Antony Hewish (b. 1924) in Cambridge, U.K. It so happened that I had an appointment with Oort in 1967 to talk about my PhD thesis research, when he had just received the publication of the discovery of pulsars. We hardly talked about my research. He was so excited and had already estimated that a neutron star would rotate with a period in the right order of magnitude if these were contracted stars. That must have been as follows (the first pulsar pulsed with a frequency of about 1.3 seconds). A neutron star has a mass of about 2 solar masses and a radius of 15 km. If it rotates around its axis in one second, its total angular moment is about of one thousandth of that of the Sun. Fair enough. By the way, this shows again how Oort worked, first make estimates to see if it is not unreasonable and only then think about the details (or leave it to others). In 1968 a pulsar was found in the Crab Nebula, which rotates even faster, namely with a period of 33 milliseconds. In 1969 it has also been discovered optically. It is assumed that supernovae and pulsars are sources of the energetic particles that we now see as cosmic rays. Without addition of new energetic electrons the synchrotron radiation of the Crab Nebula would have been extinguished a long time ago.

Gart Westerhout used the Kootwijk radio telescope to check whether the Crab was polarized in the radio radiation as well. He could not find any polarization, but that was because the beam of the telescope was much larger than the nebula, so the various directions of polarization over the image of the nebula averaged out. Walraven greatly improved the optical result by observing the Crab Nebula with a new photometer using the 80-cm telescope of the Observatoire de Haute-Provence in the French Alps (about 100 km north of Marseille). Walter Baade used the 200 in. Telescope on Palomar Mountain to take photographs of the Crab Nebula through a polaroid filter, and then rotate that filter 60° between exposures. The nebula looked very different in the different images. Oort had a student, who took it all further with a dissertation on the Crab Nebula. This was Lodewijk (Lo) Woltjer (1930–2019), son of deceased Leiden staff member Jan Woltjer. He scanned Baade's plates carefully and produced a spectacular polarization map from these (see Fig. 9.11).

Lodewijk Woltjer had first studied geology, starting in 1948, but in 1950 he had turned to astronomy. In 1957 he defended his PhD thesis on the Crab Nebula. Apart from the polarization he used spectra that Oort had obtained from Mayall of the Lick Observatory. Using these spectra he derived various properties of the Crab Nebula and modeled the synchrotron radiation. He also mapped the expansion with these spectra. I will not go into details, but mention a few outcomes. He found that the energy in relativistic electrons (which have velocities close to the speed of light) was 10^{41} joules and that the magnetic fieldstrength in the center of the nebula should be about 10^{-8} tesla. For comparison: the terrestrial magnetic field on the Earth's surface has a fieldstrength of 5×10^{-5} tesla. Woltjer found that the magnetic field in the Crab Nebula must have been caused by turbulence in the expanding nebula after the supernova explosion.

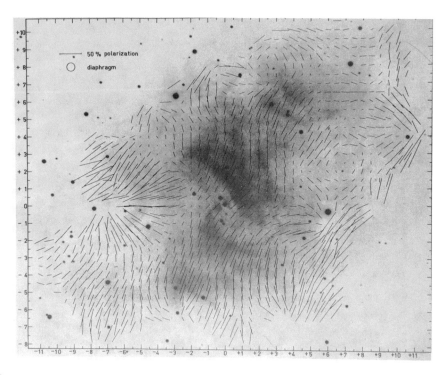

Fig. 9.11 Map of the polarization of light from the Crab Nebula, made by Lodewijk Woltjer. For this he measured photographic plates, which Walter Baade had taken with the Palomar 200 in. Telescope through a polaroid filter in three orientations, which differed $60°$ from each other. From the Oort Archives

9.5 President of the IAU

At the General Assembly of the International Astronomical Union in Zürich in 1948, Oort was relieved of his duties as General Secretary. He remained an adviser to the board. A difficult matter was that it had already been decided that the next General Assembly would be held in Leningrad (now again St. Petersburg) in 1951. It was the days of the Cold War and the Korean War, and it was not very opportune with the political situation in the USSR to organize a large meeting there then. Many Western astronomers would remain home. It was decided to move the meeting to Rome and to 1952.

Not everyone could become an individual member of the IAU at that time; it was possible since an amendment of the articles of association, but was limited to members of national and standing committees. Under Oort's secretariat he had proposed that countries could propose individuals, but that created inequality between countries. After the General Assembly of Rome in 1952, a sub-committee was established to advise on how to deal with individual

Fig. 9.12 Oort in his office in 1953. From the Oort Archives

membership, and Oort was appointed its chairman. Following the proposal that Oort's committee made, it became possible in 1955 (General Assembly in Dublin) for anyone with a PhD in astronomy to become a member, albeit on the recommendation of national committees. This fundamentally changed the character of the IAU, because now many professional astronomers became IAU members. At present there are more than thirteen thousand.

Symposia were also organized under the auspices of the IAU, the first one, on the initiative of Oort (Fig. 9.12), took place in June 1953 'On the coordination of research on our Milky Way System'. It took place at the Vosbergen Conference Centre of the University of Groningen (it no longer exists), about 10 km south of Groningen. It brought together most of the experts in the field and was a great success. An important outcome was the decision that an extensive study would be made of some areas in the sky near the center of our Galaxy, called Baade's Windows. There was very little dust there and one could look to large distances even beyond the center of the Galaxy. The most important of these areas, discovered by Walter Baade, lies on the sky only 4 degrees from the direction to the center. With the wide-angle telescope at Palomar Observatory, the 48 in. Schmidt, one would search for variable

Fig. 9.13 Some participants at the Vosbergen symposium near Groningen in 1953. On the left Bertil Lindblad en Walter Baade. In front of Oort Lukas Plaut. Behind him slightly to the right Adriaan Blaauw. From the Oort Archives

stars and expected to find many RR Lyrae stars. Of those pulsating stars one knows the absolute magnitude so that one can then determine the distance. This would enable a thorough study of the central parts of our Galaxy. Oort had a lot of interest in this program. This was finally executed by Lukas Plaut (1910–1984) in Groningen. Plaut (see Fig. 9.13) was born in Japan and grew up in Germany, where he studied astronomy. His Jewish descent forced him to come to the Netherlands, where he defended a PhD thesis in Leiden in 1939. During the war he remained in Groningen, but was arrested and taken to concentration camp Fürstenau. He survived and returned to the Kapteyn Astronomical Laboratory in Groningen.

In 1953 Pieter Oosterhoff (see Fig. 9.14) was appointed General Secretary of the IAU (after having been an assistant since 1948). He remained in this function until 1958. In that year the General Assembly was held in Moscow. Before that some wrinkles had to be smoothed because of the cancellation of

Fig. 9.14 Oort and Oosterhoff in 1950 at the Sterrewacht, probably talking about organizational or administrative issues. Pieter Oosterhoff for many years took over a large part of Oort's duties as director. From the Oort Archives

the meeting in Leningrad in 1951. Oort and Oosterhoff, who visited Russia in 1954 during the celebration of the reconstruction of the Pulkovo Observatory near Leningrad, played a significant part in this. At the end of the General Assembly in Moscow, Oort was elected President of the IAU, which would last until the next General Assembly in 1961 in Berkeley near San Francisco.

Oort's IAU presidency was marked by four important matters. First, there was China. Since its foundation in 1949, the People's Republic had regarded itself as the only China, but shortly before the General Assembly in Moscow, Taiwan had also applied for membership. The IAU Executive Board was divided along the lines of the global political situation. In addition to the President, there were six Vice-Presidents, two of them from communist countries. The six were Boris Vasilyevich Kukarkin (1909–1977) from the Sovjet Union, Bohumil

Sternberk (1897–1983) from Tjechoslowakia, Leopold (Leo) Goldberg (1913–1987) from the USA, Otto Heckmann (1901–1983) from West-Germany, Robert Methven Petrie (1906–1966) from Canada and Richard Hugh Stoy (1910–1994) from South-Africa. The General Secretary was Donald Harry Sadler (1908–1987) from the U.K. (he was superintendent of the Nautical Almanac Office of the Royal Greenwich Observatory). Kukarkin and Steinberk felt that the application should be rejected, with the argument that astronomy was hardly practiced in Taiwan. Oort, however, emphatically felt that this should not be an argument, because there was nothing about it in the statutes. It was a delicate issue, but after much discussion, the board decided with five to two to admit Taiwan. The People's Republic of China immediately withdrew. It was not until 1978, at the General Assembly in Montreal under the Presidency of Adriaan Blaauw, that both China's became members of the IAU and in 2015 the General Assembly was held in Beijing.

Another difficult subject was raised by two young French astronomers, Jean-Claude Pecker (1923–2020) and Evry Schatzman (1920–2010), who were dissatisfied with the functioning of the union. According to them, the committees became too big and too little happened between the General Assemblies. They proposed to make a two-dimensional division of commissions, on the one hand on the basis of subjects of study (planets, stars, etc.) and on the other hand techniques, methods and instrumentation (astrometry, spectroscopy, and so on). Oort took this very seriously, and together with Sadler and Oosterhoff he had various conversations with them. In the end, the French partly got their way; each committee would have a small Organizing Committee that coordinated matters between General Assemblies (such as the choice of topics for symposia) and for their part of astronomy set up the program during the General Assemblies. At the next General Assembly, Pecker became Assistant General Secretary, to take over the position three years later.

All this required a lot of tact from Oort. More difficult was the commotion around Project Westford. The US Army wanted to use a satellite to take a large number (480 million) of copper needles (1.78 cm by 25.4 μ)^2 into space, which would then spread and could be used for telecommunication. The length of the needles was the wavelength of the radio radiation used by the communications station of the Massachusetts Institute of Technology MIT, Cambridge, Massachusetts, which was located near Westford in the same state. This was a serious threat to observatories. Despite many protests from the IAU and pressure on governments around the world, the project went ahead. The launch took place in 1961, but the needles did not spread. Unfortunately it was not canceled then and was tried again in 1963. At that time it succeeded,

---

^{2}Micron or μ is one thousandth of a millimeter.

but the quality of the communication turned out to be too poor. Fortunately, the effects on the astronomical observations turned out to be minimal, but it showed how weak astronomy's position was during the Cold War (and still is) in matters of military (and nowadays commercial) interests.

But the most difficult issue, which took Oort an enormous amount of time, was that of the protection of radio bands for astronomy. Telecommunication, especially after the launch of Sputnik I in 1957, was potentially a huge problem for radio astronomy, which could not use interfering signals. As General Secretary, Oort had already had dealings with the ICSU, the International Council of Scientific Unions, and he still had good contacts there. In 1958, the ICSU was setting up a committee to safeguard the interests of scientific space research, including the exchange of scientific information. This was called the Committee on Space Research (COSPAR). Oort was unable to attend the first meeting and asked Henk van de Hulst to do so on his behalf. To his dismay, van de Hulst returned to Leiden having been elected the first chairman.

Now there was another body, established in 1865 as the International Telegraph Union, which had become an agency under the United Nations in 1947 and was now called the International Telecommunication Union UTI (abbreviation from the French), and which coordinated the worldwide use of telecommunications. In 1959, this institution held a conference in Geneva in order to decide on the distribution of frequencies for radio stations and telecommunication. Oort realized that it was extremely important for radio astronomers to be present there in order to protect the important frequencies, such as that of the 21-cm line. It lasted from mid-August to mid-December (more than 6000 requests had been filed) and Oort felt that astronomy should be present there all that time. Together with the General Secretary of the IAU, Sadler, he arranged this to happen. He was also there himself for some time and gave a public lecture, because no one at the conference had any notion of radio astronomy. The lecture was also published in the well-read American periodical, the *American Scientist*. Eventually, radio astronomy was assigned some bands as protected frequencies, including the 'unconditional use of the 1400–1427 Mc/s band' for the 21-cm line. This is for a very important part due to Oort's efforts.

In order to permanently defend the protection of the radio astronomy and space research bands, Oort ensured that the IAU, together with the International Union for Radio Science URSI, abbreviation again from the French, and COSPAR established and supported a standing committee under the name Inter-Union Committee on Allocations of Frequencies IUCAF. It still exists under that acronym, although the name has changed to Scientific Committee on Frequency Allocations for Radio Astronomy and Space Science.

Fig. 9.15 Opening ceremony of the IAU General Assembly in Berkeley in 1961. From the left Ira Bowen, Oort, Adlai Stevensen, Leo Goldberg and Donald H. McLaughlin. See the text for details. From the Oort Archives

The next General Assembly of the IAU took place in Berkeley in 1961 and Oort was President. Fig. 9.15 shows Oort during the speeches at the opening ceremony. Speaking is Leo Goldberg, chairman of the American National Committee. Adlai Ewing Stevenson (1900–1965), to the right of Oort, was the American Ambassador to the United Nations. To the far right Donald Hamilton McLaughlin (1891–1984), Regent of the University of California. And on the left Ira Sprague Bowen (1898–1973), director of Mount Wilson and Palomar Observatories. Oort first spoke in English and then summarized his speech in French as the second official language of the IAU. That summary was not much shorter than the English original. He complimented the University of California with the recently completed (1959) 120 in. Telescope on Lick Observatory (later named after Donald Shane—Fig. 9.16—who was then director of Lick) and honored George Hale with the Mount Wilson and Palomar Observatories, but spoke extensively about radio astronomy. After a brief reference to frequency protection and Project Westford, he emphasized the great potential of space research. He also spoke about his concern about the growing size of the IAU. He referred to the fact that he was probably the only astronomer who had attended all the IAU General Assemblies since

Fig. 9.16 Oort with Donald Shane and his wife Mary Lea Heger in 1961 at IAU General Assembly in Berkeley. From the Oort Archives

Rome 1922, and who had witnessed the growth at close quarters. Incidentally, he continued to attend those meetings until the 17th in 1979 in Montreal. In 1982 he chose to attend some other symposia around that time.

9.6   The Dwingeloo Radio Telescope

In 1950 the provisional ZWO had become formal. In that year it had a budget of 1 million guilders; the intention was that this would grow to 5 million in 1955, after which it would go down again and ZWO would be disbanded. The reconstruction of scientific facilities for pure research in Netherlands would have been brought back up to standard. This never happened; ZWO continued to exist and was converted in 1989 to NWO, the Netherlands Organization for Scientific Research, which now funds the majority of research at universities and research institutes. When the 21-cm line was discovered, the board of the Foundation RZM decided to submit a new application for a 25 m radio telescope. Now that the line had been discovered, ZWO was willing to support it. Meanwhile, estimates of the costs had risen to half a million, mainly due to new design by Werkspoor.

The construction was immediately taken in hand. Crucial to the success was the availability of two engineers. Lex Muller was responsible for the electronics as in Kootwijk. The other was Bernard G. (Ben) Hooghoudt (1924–1995), responsible for the mechanical structure. Hooghoudt was not employed by RZM, but by PTT and RZM refunded his personnel costs for his work to the PTT. Hooghoudt was especially important as an intermediary between the Foundation and Werkspoor, since he kept the cost increases under control. Hooghoudt also had an essential influence on the design. His experience in Dwingeloo made him a world expert on the mechanical structure of telescopes and he contributed to many telescope designs around the world, especially (but not exclusively) for radio and millimeter wavelengths.

Henk Bannier, the director of ZWO, was also essential. In principle, he attended all board meetings of RZM and was therefore permanently fully informed of all matters. This put him in a position to inform the board of ZWO with authority on all aspects, including delays and extra costs. Oort and Bannier also had an excellent relationship on a personal level. The total investment in cash (i.e. not in manpower) ultimately amounted to 0.6 million guilders (1.9 million € in current purchasing power).

The design of the radio telescope was what is called alt-az, derived from altitude (height) and azimuth. This means that the telescope could rotate around a vertical and a horizontal axis. This made the structure as simple and robust as possible. However, it complicates the determination of a position of an object in the sky in astronomical coordinates in relation to these two axes for a certain point in time and then in order to track an object. Nowadays all telescopes are alt-az, but they are also computer controlled and then it is no problem. In Dwingeloo an ingenious device had been developed, which made this conversion mechanically; a kind of analogue computer, which was called the 'pilot'. Also the orientation in the focal plane changes when an object is tracked and in an optical telescope the plate holder or instrument has to rotate. In Dwingeloo astronomers wanted to measure the polarization of the radio radiation; this is done with two perpendicular dipoles which detect the radiation. Their orientation in relation to the sky changes in an alt-az telescope and this must be taken into account when processing the observations.

The 'mirror' of 25 m diameter was covered with a fine mesh. Those meshes must be significantly smaller than the wavelength of the radiation in order for it to function as a reflecting surface; in Dwingeloo those meshes were 15 × 15 mm. The 'beam', the smallest detail that can be seen, depends on the wavelength and diameter of the mirror as a result of imaging at the focal point, where radiation from all parts of the mirror surface interferes with each other. For Dwingeloo at 21 cm that was a diameter of 34 arc min, just over half a degree or roughly the Moon in the sky. This size corresponds to where the sensitivity of the telescope is half of what it is precisely on the axis, the direction in which it is pointing.

The selection of the site where the telescope should be located was not trivial. There must be no car traffic nearby, because the electric sparks in the ignition of an engine cause disturbing radio radiation. In the end, the province of Drenthe was the most obvious. But local authorities were apprehensive about the negative impact on the tourist industry if cars would be banned in a fairly large area around the installation. Farmers and nature conservationists were also reluctant. Here, Bannier played a crucial role in consultation with local authorities, the Forestry Commission and the Ministry of Fishery and Agriculture to obtain the necessary permits. Eventually a location was chosen near Dwingeloo, in the hamlet of Lhee, in a nature reserve, the 'Kraloër and Dwingeloosche Heide', now the most important part of the National Park Dwingelderveld. The telescope was erected on the northern edge of a large heath-land area, so that there was an unobstructed view towards the southern horizon and the buildings could be placed between the trees of the adjoining woods (see Figs. 9.17 and 9.18).

The inauguration of the telescope was done by Queen Juliana on 17 April 1956 (see Fig. 9.19). Prior to this there had been a get-together in Dwingeloo, where Oort was awarded a Royal Decoration by the Minister of Education, Arts and Sciences in the presence of the Queen. The minister was Joseph Maria Laurens Theo Cals (1914–1971). The decoration was high, 'Knight in the Order of the Dutch Lion', the lowest rank in the highest civil order.

Fig. 9.17 Aerial view of the Dwingeloo Radio Telescope around 1960 with Würzburg antennas on both sides, which were mainly used for solar research. The associated buildings are located in the upper right-hand hidden in the forest as stipulated when the permit was issued

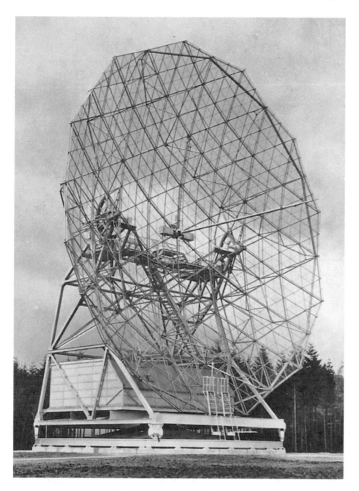

Fig. 9.18 The 25 m telescope in Dwingeloo not long after construction. The long mast reaches to the focal point, where the dipoles are placed to detect the radio radiation. After some time this was replaced by a box with the first part of the receiver and amplifier, which required a more solid construction. The mast was then replaced by a tripod, and later by a quadropod. From the Archives of the Sterrewacht Leiden

Distinctions in this order are rare and awarded only to special persons. For example in the annual decoration 'shower' at the occasion of the King's or Queen's birthday no more than ten or twenty out of three thousand are in this high order. The rest is in the order of Orange-Nassau. The opening included a push by the Queen on a button, following which the telescope would start moving towards a preselected position. However, the control was still by hand, so a light went on in the control room. Due to a mistake the telescope first

Fig. 9.19 Queen Juliana and Oort after the inauguration ceremony of the Dwingeloo Radio Telescope on April 17, 1956. From the Oort Archives

moved in the wrong direction, but Gart Westerhout could intervene only just in time [102].

Herman Kleibrink, commissioned by RZM, made a movie about the construction of the Dwingeloo Radio Telescope, which is available on the Web [103].

The question is sometimes raised whether this investment for astronomy in the post-war period, when a lot of funding was needed to rebuild the country, was really justified. Some, including Adriaan Blaauw in an interview with Jet

Katgert-Merkelijn in the 1990s, were skeptical about this. Information concerning this is available in the extensive historiography of ZWO [104]. We then see that the total budget of ZWO for the decade 1951 up to and including 1960 was 51.6 million guilders. About half of these went to the FOM, the Foundation Fundamental Research on Matter—a cuckoo's egg in the ZWO nest, according to this author. FOM did nuclear physics and mass spectroscopy (uranium enrichment), both important in the context of nuclear energy. Including the operating budget (365,000 guilders in 1959), RZM received no more than 5 percent of this over this period. This is not an exorbitant amount at all, especially if we consider that it is 5 percent of the budget for *pure* scientific research.

Fig. 10.0 Oort at his desk in 1955. From the Oort Archives

10

The Structure of Our Galaxy and Dwingeloo

... an autocratic gentleman of about seventy,
with a balding skull and a sharply etched face.
[...] That was the director, he said [...],
– who had not only shown that the Milky Way system was rotating,
but also that it had spiral structure.
Harry Kurt Victor Mulisch (1927–2010)

With their radio telescopes they can capture wisps of radiation
so preposterously faint that the total amount of energy collected from outside
the Solar System from all of them together since collecting began (in 1951)
is less than the energy of a single snowflake striking the ground.
Carl Edward Sagan (1934–1996)

10.1 Administrative Work

The professors of a university were expected to take on administrative
functions by turns. The highest body in the organization was the Board of
Curators (Trustees) with next to it the Senate, the collection of all professors

---

Harry Mulisch was a prominent Dutch writer. The quote (my translation) is from his novel *The
discovery of Heaven* [105]. One of the main characters, Max Delius, is—just as I was—a Ph.D.
student in Leiden around 1970, studying the Galactic center in Dwingeloo and preparing to use
the Westerbork Telescope; just like him I was offered a Carnegie Fellowship at Mount Wilson and
Palomar.
The numerical justification of the second quote is questionable, see [106]. The comparison is sometimes incorrectly attributed to William McGuire (Bill) Bryson (b. 1951), who actually does quote it
but with proper reference in *A short history of nearly everything* [107].

© The Editor(s) (if applicable) and The Author(s), under exclusive license
to Springer Nature Switzerland AG 2021
P. C. van der Kruit, *Master of Galactic Astronomy: A Biography of Jan Hendrik Oort,*
Springer Biographies, https://doi.org/10.1007/978-3-030-55548-1_10

(for a description of the history of Leiden University see [108]). Decisions about budgets and appointments were made in the Senate. It was chaired by the Rector Magnificus, assisted by a Secretary (the next Rector) and a Praesidium Senatus, consisting of representatives, usually Deans of the Faculties. The Rector had a term of office of one year, Deans two years, the latter assisted by a Board of Professors, half of whom was replaced by others every year. The most senior professor was expected to assume the position of Rector for one term, although some accepted a term several times.

Oort was Dean of the Faculty of Mathematics and Natural Sciences from 1955 to 1957. During that period serious budget cuts were made. Reconstruction of the country after the Second World War had resulted in an economic boom. However, the growth of the economy began to show signs of slowing down and cutbacks seemed inevitable. The Minister of Education, Arts and Sciences, Jozef Cals, wanted to appoint a committee to advise the universities on how to deal with this.

Oort, however, felt that this was far too slow and that for the exact sciences the situation was exceptionally serious. He therefore decided to approach the Prime Minister himself, Willem Drees (1886–1988), directly. He sent a draft letter to his fellow Deans at the other two state-funded universities in Utrecht and Groningen (in Utrecht this happened to be astronomer Marcel Minnaert) and he obtained their agreement and support. The final draft was presented to the Deans of the Municipal University of Amsterdam and the Free University, but they were left little time to comment. Oort sent the letter to Prime Minister Drees on December 7, 1956. The letter was seven pages long, including two appendices containing tabular and other detailed information that Minnaert had collected. It was emphasized in the letter that there was an imminent and critical need to quickly appoint more manpower and build more laboratories, as science and technology were in danger of becoming a bottleneck for progress. The Netherlands as a whole would be in danger of lagging behind the United States. The budgets for their Faculties would have to grow, but a growth of 6 percent (which had proposed before the impending budget cuts) was in fact even insufficient. This pressing letter had important consequences.

A meeting of some Deans and Secretaries of Faculties with Minister Cals took place in April 1957. The minister apparently did not feel passed over. The Deans urged that an advisory committee be set up. Also present was the highest civil servant in the ministry, Arie Johannes Piekaar (1910–1990). He had studied indology, and after a diplomatic career in the Dutch East-Indies, among other places, he worked at the ministry mainly on science policy, in which function he has been of great significance, until his retirement in 1975. The Minister wrote a letter to Oort and his colleagues announcing that he would do everything in his power to prevent the threat of a staff freeze in their faculties and to find funds for new buildings. And he asked for a proposal for the advisory committee and its members.

Oort sent a proposal for a committee, with names of possible members, and the committee was installed on December 3, 1957. It was recognized under the name of its chairman, physicist Hendrik Casimir, who as we saw was director of Philips NatLab. Other members were Oort, plant physiologist Willem Hendrik Arisz (1888–1975) from Groningen, physicist Cornelis Gorter from Leiden, physical chemist Jan Theodoor Gerard Overbeek (1911–2007) from Utrecht and Hendrik Willem Slotboom (1904–1996), chemical engineer at Royal Shell Netherlands. The committee reported back in October 1958 (see also [109]).

The increase in students in the natural sciences as a result of the post-war birth wave was even faster than expected; it grew by a factor of three or so between 1957 and 1970. The budgets for scientific education grew by a factor of six. Much of this was due to the Casimir Committee. There was no exemption from military service for the most promising students, as the Committee had proposed; there was, however, a postponement possible for six or seven years until the doctoral exam (masters) and another four years for a PhD. Oort's idiosyncratic and unconventional action had paid off and had resulted in clear and essential consequences for the practice of natural sciences in the Netherlands.

When it came to Oort's turn to become Rector, he resisted strongly. He wrote to the incumbent Rector:

> ... in practice, the time needed to properly fulfill the Rectorate will be taken away from the time I have left for my own scientific work. Now it is already in a sad state anyway. [...] I find this an unsatisfactory state of affairs. For my talent lies first and foremost in the direction of my own research. There are such wonderful things in the Universe to be found at the moment that I cannot force myself to decide to turn away from them, nor can I justify it.

The discussion as to whether it was really such a good idea that everyone would be Rector Magnifiucus for a year in turns continued for a while, but eventually the system was abandoned. Oort was never Rector Magnificus.

Oort wanted the *ad hoc* Casimir Committee to be succeeded by a permanent committee. A new law on higher education in 1960 provided for the establishment of an Academic Council. Oort turned down the invitation to be part of the sub-committee for the exact sciences, but for its first year he acted as an interim chairman.

Oort has also been active in administrative work outside the bodies directly related to the university. For example, for some time he was a member of the Advisory Board of the Boerhaave National Museum for the History of Natural

Sciences. He was also a board member of the Leiden Academic Arts Centre LAK, where he and Mrs. Oort were active in the visual arts. Their contacts with Hendricus Bremmer will have stimulated this. They were involved in the organization of an exhibition *From Fantin Latour to Picasso* in 1950, in which several paintings from the collection of the Kröller–Müller Museum were exhibited in Museum de Lakenhal (Dutch for Cloth Hall) in Leiden. An attempt to realize a review exhibition of the Leiden painter Jan Josephszoon van Goyen (1596-1656) failed, to the frustration of Oort and his wife due to the lack of interest by the museum.

10.2   Stellar Populations at the Vatican

For a long time the Catholic Church has had its own observatory. This goes back to a commission that advised Pope Gregory XIII (Ugo Buoncompagni, 1502–1585) resulting in the famous revision of the calendar in 1582. The observatory, now officially called Specula Vaticana, existed as an institute in a number of forms since 1774. Since 1891, it has been located in the Vatican itself. The Dutch Jesuit and astronomer Johannes Wilhelmus Jacobus Antonius Stein S.J. (1871–1951), who had a PhD from Leiden in 1901, became its director in 1930, and he was responsible for the relocation to the Pope's summer residence and vacation retreat at Castel Gandolfo because of the city lights of Rome. Today, in addition to this location, the most important observing facilities are located near Tucson in Arizona. In addition to the calendar, the papal interest was also prompted by the importance of astronomical developments, which provided changing insights into our world view.

Now, in 1944, Walter Baade had presented the concept of Stellar Populations, which gave a whole new perspective on the structure of galaxies and their formation and evolution. As a German citizen during the Second World War, Baade was restricted to an area around Pasadena, but he had permission to go to Mount Wilson. Since the American astronomers were for a large part called upon to contribute to (mostly technical) developments in connection with the war, he had a lot of observing time at his disposal. In addition, the sky was unusually dark in Los Angeles due to black-outs in view of the threat of Japanese attacks. On plates of the Andromeda Nebula (see Fig. 6.4) he had noticed that there was a chance that he could see the brightest stars in the central parts individually. This was of course possible in the outer parts, because Hubble had already used Cepheids there in the 1920s to determine the distance. Cepheids are pulsating stars, with well-defined shapes of brightness profiles and a very regular period of brightness variations. They are named after the prototype δ

Cephei and are relatively massive, and therefore relatively bright, stars in the late stages of evolution, with a range of luminosities or absolute magnitudes and the period of the pulsation correlates very accurately with that period. RR Lyrae stars, that we saw were used to find distances of globular clusters, are similar objects, but now in an advanced stage (helium-burning) of stars of moderate mass. Since Cepheids are luminous they can easily be seen in outer parts of the Andromeda Nebula and other nearby spiral galaxies.

But nothing was known about what the brightest stars in the central parts were. Eventually Baade succeeded when he used new photographic emulsions (of the Kodak Company), which were, contrary to what was available before, sensitive to red light. While in the vicinity of the Sun the brightest stars were very young and hot (i.e. blue) *O* and *B* stars, in those central parts those were absent and it were the red giants that dominated. Star formation no longer existed there. Baade deduced from this that there are two Stellar Populations, which he called Population I and II (later these became also designated as the disk and halo population). In the relatively flat disks of spiral galaxies, in which star formation apparently takes place at this time (in the spiral arms), we find Population I with stars of all ages among which blue, massive and young stars. The halo, arranged in a more spherical but centrally concentrated structure, contains Population II, which is made up only of old stars, of which the globular star clusters are the most prominent parts.

Eventually, this led to a doubling of the distance scale in the Universe and therefore of its age. It turned out that the pulsating stars used by Hubble to determine the distance of the Andromeda Nebula belonged to that halo population II (and are now called W Virginis stars) while their absolute brightness was calibrated with the Cepheids in the Solar neighborhood, which belong to Population I. There is a difference in their luminosity, so that distances obtained in this manner turned out to be underestimated by a factor two.

Meanwhile, much more had become known about the evolution of stars and their properties. Since the work of Hans Albrecht Bethe (1906–2005) and Carl Friedrich von Weizsäcker (1912–2007) just before the war, energy production was understood as the result of nuclear reactions, in which hydrogen is converted into helium. Theoretical work of Subrahmanyan Chandrasekhar (1910–1995) in Cambridge and Yerkes, Martin Schwarzschild and Fred Hoyle (1915–2001) in Princeton and Cambridge, and observations by Allan Rex Sandage (1926–2010) of globular clusters with the new 200 in. on Palomar Mountain, had resulted in tremendous progress in our understanding of the late stages of stellar evolution. Also, the work of American astronomer Nancy Grace Roman (1925–2018) had made it clear that a fundamental difference between the stars in the two populations was that the halo population (Population II) contained hardly any elements heavier than helium, while for stars in the disk population (Population I) this could be as much as a few percent (in the Sun this is 2%). Hydrogen and helium always occur in a ratio of three to one (in mass), in all stars in both populations. The insight emerged that in the early Universe there was practically only hydrogen and helium and that elements such as carbon and heavier were all formed in stars and dispersed in the interstellar medium during supernova explosions. This theory of nucleosynthesis in stars is, in my opinion, perhaps the greatest achievement of natural science in the twentieth century. It was summarized in 1957 in a famous, long and widely-read review article by two young British astrophysicists, Eleanor Margaret Burbidge (née Peachey; 1919–2020), Geoffrey Ronald Burbidge (1925–2010), nuclear physicist William Alfred (Willie) Fowle (1911-1995), and Fred Hoyle, often referred to as B^{2}FH.

Fig. 10.1 Walter Baade (right), his wife Johanna (Hanni) Bohlmann and Martin Schwarzschild in 1956 at an unspecified location. From the Oort Archives

Immediately after the war, the director of the Vatican Observatory, Daniel Joseph Kelly O'Connell S.J. (1896–1982), conceived a plan to organize a meeting at which only a small number of leading scientists would meet for a week to synthesize everything known into an general understanding of the structure, formation, and evolution of galaxies. The idea of such a *Semaine d'Étude* came before the war from the then director Johannes Stein. In the meantime, a few other such study weeks had been held on non-astronomical subjects, under the auspices of the Pontifical Academy of Sciences. Jason John Nassau (1893–1965), an astronomer who had done work (stimulated by Baade himself) on stellar populations in the form of the distributions of samples of red stars, had suggested this subject for the Semaine to O'Connell. Oort, Nassau, Schwarzschild and Baade (Fig. 10.1) in particular were of course strongly involved in the organization.

The group photograph of the participants is shown in two parts in Fig. 10.2. In the upper part are Oort and Walter Baade in front with Lindblad to the left of Oort. On the far left is Nassau. He and Baade look somewhat alike and they cultivated that. In many conference photographs they wear similar clothes and adopt a similar pose. Here this is the case to a lesser extent. Next to Nassau stands Daniel Chalonge (1895–1977) of the l'Observatoire de Paris, specialized in spectral classification of stars. Next Martin Schwarzschild, best known for being the first to calculate the evolution of a star after the Main Sequence phase along the giant branch (towards the phase of red giants) in the Hertzsprung–Russell diagram. Next to him Adriaan Blaauw and to his right Otto Heckmann from Hamburg, who worked in cosmology and the application of the theory of general relativity. In the background from the left Edwin Ernest Salpeter (1924–2008) from Cornell University, an expert in the field of energy production in stars, as well as Bengt Strömgren and Lyman Spitzer.

The lower panel shows from the front left Father O'Connell, Giuseppe Armellini (1887–1958) and famous cosmologist Father Georges Henri Joseph Édouard Lemaître (1894–1966) from Belgium. Arminelli and Hermann Brück (just outside the figure) were astronomers who were present as members of the Pontifical Academy. In the back row Fred Hoyle (Cambridge), who was the first to show how helium can be converted into carbon in red giants during nuclear reactions, William Wilson Morgan (1906–1994) of Yerkes Observatory, who investigated the distribution of OB stars and star formation in the solar neighborhood, and George Howard Herbig (1920–2013) of Lick Observatory, an expert on young stars. On the far right Allan Sandage of Mount Wilson and Palomar Observatories, who discussed his work on stars in the neighborhood of the Sun, in globular clusters and in nearby galaxies. In the back

Fig. 10.2 Participants of the *Semaine d'Étude* in 1957 at the Vatican on Stellar Populations. For the sake of clarity, the original photograph has been cut in two (and some Vatican staff members on the sides have been omitted). Persons are identified in the text. From the Oort Archives

Willie Fowler from Caltech, who experimentally studied atomic nuclei to learn more about nuclear reactions occurring in stars, and Andrew David Thackeray (1910–1978) from Radcliffe Observatory, who worked on populations in the Magellanic Clouds.

This historical study week provided a picture of the formation and evolution of our Galaxy. It gave the concept of Stellar Populations a more solid foundation, enabling it to play the defining role it has had in astronomy in the second half of the twentieth century. It emerged very well in the summary of Oort at the end of the Semaine (Hoyle summed up the theoretical work). The starting point is that the early Universe consisted of three-quarters hydrogen and one-quarter helium (there have been traces of lithium and some of the lightest elements). This is when the first stars came into being, of which the very first ones were probably heavy, so there are none left. These synthesized the first carbon, oxygen, nitrogen, and heavier elements that were added to the gas by supernova explosions. Later generations can be seen in the halo populations and the globular clusters. These still contain very few heavier elements. These components of galaxies consist only of old stars and are spatially arranged in a more or less spherical structure that rotates little, but in which high relative velocities occur. This is Population II. The disk population later formed from clouds of gas, in which the rotation due to the contraction increased. Due to the collisions between gas clouds, which are now enriched with heavier elements, the mutual velocities decreased considerably. This became a flat disk and formed Population I, in which star formation still is proceeding.

Fig. 10.3 The Oorts and Walter Baade relax a few days after the Vatican symposium on the Italian coast at Amalfi. From the Blaauw Archives

The Vatican summary shows a series of populations, in which the old disk has the central position in the sequence. At one end is the halo Population II and opposite on the other end the extreme Population I of very young stars (and gas). This picture is still valid, broadly speaking, although the formation of the halo population in particular is now seen more as the result of the merging of small satellite systems especially early on in the evolution. A major result after which some relaxation was well deserved (Fig. 10.3).

10.3   South-Africa and Chile

In 1950, Fjeda Walraven had taken over the position of observer from Leiden in South-Africa from Adriaan Wesselink, who had moved to Radcliffe Observatory. The light pollution of Johannesburg was still a problem. Walraven and his wife Johanna Helena (Jo) Verlinden (1921–1989) worked usually together on observational programs. However, in 1955 the Walravens had returned to Leiden, where Fjeda Walraven and Oort did the work on the polarization of the Crab Nebula. The new Leiden observer was Andreas Bernardus (André) Muller (1918–2006), who had written a dissertation under Oosterhoff. Muller would play a major role in the development of the European observatory in the southern hemisphere.

This history begins with the wish to move the southern station of the Sterrewacht Leiden to a darker location than Johannesburg. One possibility for this was the southern (Boyden) station of Harvard College Observatory, which was located at Mazelpoort, 20 km from Bloemfontein in Orange Free State. This is almost 400 km to the southwest from Johannesburg. However, Harvard University in 1953 announced its intention to close the facility.

In the spring of 1953 Baade visited Leiden for a few months. Oort had invited him for more than three months, but because Baade's visa application (an entry visa to return to the United States) had been lost, it was shortened. It was just before the Vosbergen Conference, which has already been described, and Oort and Baade had had many long talks about both the preparations for that conference and the future of European astronomy. Their conclusion was that Europe had to build a large observatory in the southern hemisphere. Baade described his idea of its instrumentation as consisting of a 120 in. Telescope as at Lick Observatory, and a 48 in. Schmidt telescope as at Palomar Observatory. The infrastructure at Boyden Station near Mazelpoort seemed an excellent opportunity to take as a starting point; cooperation might convince Harvard to perhaps continue their operations there.

Oort called on colleagues from European countries to talk about this matter in Leiden on the day before the Vosbergen Conference, Sunday June 21,

1953. These were Paul Eugène-Edouard Bourgeois (1898–1974) representing Belgium, for France André-Louis Danjon (1890–1967), for (West-)Germany Otto Heckmann, and Bertil Lindblad for Sweden. Harold Spencer Jones of the U.K. was unable to attend, but Oort updated him during the Vosbergen meeting. Adriaan Blaauw was present as well; however, he would move to Yerkes later that year and for the time being play no part in the plan. An important guest was Henk Bannier, who was a member of the Council of CERN, the European organization for nuclear research (CERN actually stands for 'Conseil Européen pour la Recherche Nucléaire'). The European Southern Observatory ESO would be modeled on this. The investment was estimated at 2.5 million $ and the operation would grow to 100,000$ per year (20 million and 0.8 million € respectively at present).

After this initial exploration, in which the participants in the Vosbergen conference met extensively during a boat trip on the IJsselmeer (see Fig. 10.4), enthusiasm quickly grew. On January 25–27, 1954 representatives of Belgium, France, (West-)Germany, the U.K., the Netherlands and Sweden met in Leiden in the Senate Room of the Academy Building. At the end of their talks they signed a document, in which the intention was laid down to create a European observatory in the southern hemisphere (see Fig. 10.5). This marks the beginning of ESO. A copy of the document now hangs at the entrance of the Senate Room. Two persons who were present but did not sign, but who turned out to be essential, were Bannier and his Swedish opposite number Gösta Werner Funke (1906–1991) (See Fig. 10.6). They knew each other from the CERN Council and were well acquainted with the European diplomatic world as far as scientific financing was concerned.

The development of the plan was as is usually the case a story of ups and downs (more downs than ups), but Oort and his collaborators persisted. Soon the United Kingdom decided to leave the consortium and opt for the plan to build together with Australia the 3.9 m Anglo-Australian Telescope at their Siding Spring Observatory in the framework of the British Commonwealth. After a few years, Oort was elected permanent President of the ESO Council. Oort tried to obtain funding from the Ford Foundation. This was founded in 1936 by Edsel Bryant Ford (1893–1943) and Henry Ford (1863–1947) to promote international cooperation. In 1959, a million dollars was allocated. At that time this was one fifth of the total planned investment and the condition of the Ford Foundation was that before actually making this available four of the five countries involved had to commit to their share; if that was met more or less the full amount of funds would be available. The text of a convention, modeled on that of CERN, had now been drafted and the Ford Foundation's commitment gave the final push to formally start. On October 5, 1962, the

Fig. 10.4 Photographs taken during a boat trip on the IJsselmeer with several participants at the Vosbergen Conference in 1953. In the top picture Mrs. Oort with Strömgren en Lindblad. At the bottom Oort and Harold Spencer Jones, probably making plans for a European southern observatory. From the Blaauw Archives

convention was signed. The required ratification by the national parliaments was already completed in 1963 by that of the Netherlands, Sweden and West-Germany, and France followed in January 1964. This was four of the five and this freed up the funds of the Ford Foundation. Belgium only ratified in 1967; a few months earlier Denmark had joined as the sixth member.

In the meantime much work had been done in South-Africa to locate a suitable site for the observatory. It was soon realized that there were better places than Mazelspoort and the search moved to the southwest into the Karoo. The Netherlands and Oort did not play a major role in this. The Unie Sterrewag decided to move to Hartebeespoortdam near Pretoria (about 50 km northeast of Johannesburg) and in 1954 the Franklin-Adams Telescope and

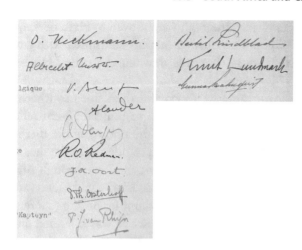

Fig. 10.5 Signatures under the ESO declaration of January 26, 1954. Signatories are Bourgeois (Brussels), Couder en Danjon (Paris), Roderick Oliver Redman (1905–1975) (Cambridge), Oort, Oosterhoff (both Leiden), van Rhijn (Groningen), Lindblad (Stockholm), Lundmark (Lund) and Karl Gunnar Malmquist (1893–1982) (Uppsala). From the Oort Archives (at ESO)

Fig. 10.6 Two important diplomatic players in the creation of ESO, Henk Bannier (left) and Gösta Funke, directors of the research councils of the Netherlands and Sweden. Photo taken in 1967 at CERN [110]

the Leiden Rockefeller Astrograph were moved to Hartebeespoortdam. André Muller moved along. The Boyden Station continued with funding from an international consortium, but was transferred to the University of Orange Free State in 1974.

There was also an initiative to expand the Leiden southern station with a larger telescope. Oort presented a plan to build a 90-cm reflector telescope for photoelectric photometry of stars; the photographic method as with the Rockefeller was no longer up to modern standards. In the meantime Oort had gained experience with the photoelectric method through his work in 1948 with Hiltner of Yerkes Observatory at the McDonald Observatory. Yerkes had developed a plan to build a 100 in. telescope according to a new concept (this was never realized) and the Leiden telescope was intended a smaller version of this. Yerkes was prepared to supply the optical components at the cost price. The optics were eventually financed by ZWO, and the telescope and housing by the University of Leiden; the total cost now translated into just under 100,000 € at present rates. It will have come in handy that former astronomy PhD student Jaap Baron de Vos van Steenwijk was both president of the university (of Curators) and at the same time chairman of the Sterrewacht Fund. This fund was the actual owner of the Rockefeller Astrograph. As we have seen, Fjeda Walraven had built the photometer for the Zunderman telescope in Leiden and was now expected to build a photometer for the new telescope in Hartebeespoortdam as well. He and his wife went back to South Africa. André Muller had also come to Hartebeespoortdam to put the Rockefeller Astrograph back into use.

The new telescope was called Light Collector and was equipped with a photometer that could simultaneously measure in five wavelength regions (Fig. 10.7). It came into operation in 1958 and much important work was done with it, especially on pulsating stars (RR Lyrae stars). In 1978 the telescope has been transferred to Chile and operated for a long time at ESO's La Silla Observatory

Both Muller and Walraven worked at the Hartebeespoortdam site and one of the two had to return to Leiden. They did not get along very well. Oort chose to leave Walraven in South Africa, because he expected more from him with the new facilities than from Muller. The latter came back to Leiden with his family, but was told that there was no work and no appointment for him. Meanwhile, Adriaan Blaauw had left for Yerkes Observatory in the autumn of 1953. When in 1957 Pieter van Rhijn reached the age of 65 and was to retire, Oort tried to lure Blaauw into moving back to the Netherlands and take over the directorship in Groningen. Because of the new possibilities of radio astronomy, Blaauw did like the idea, but at first he stipulated an increase in staff of the Kapteyn Laboratory, which he was granted by the University

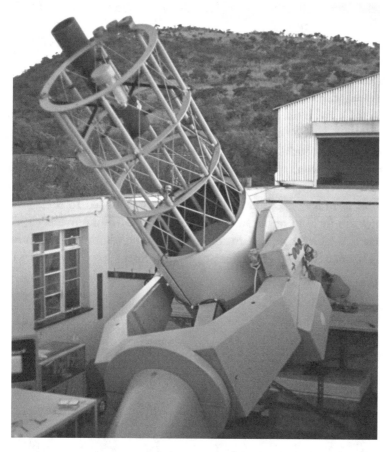

ig. 10.7 The 90-cm Light Collector at the southern station of the Sterrewacht Leiden at Hartebeespoortdam (later transferred to ESO at La Silla). The photometer is located under and to the side of the telescope. The telescope was unique in the extremely fast and accurate way it was able to point at a position in the sky using predefined coordinates. Photograph made available by Jan Willem Pel

f Groningen. When he heard about André Muller's situation—he disagreed trongly with the way Oort had treated him—he offered Muller a position in Groningen.

The search for a good site in South-Africa for the ESO observatory had not yet yielded any excellent locations. Meanwhile, a consortium of American universities was making plans to build an observatory in Chile. In 1960 Oort, who was aware of this, had asked Donald Shane of Lick Observatory for more information on this and Shane gave that to him; he liked the idea, because he hoped that cooperation between Europe and the United States, albeit each with their own telescopes in relatively nearby locations, would be attractive to all. A team under Jürgen Stock (1923–2004), a student of Otto Heckmann, who worked in Chicago, had carefully searched in Chile and on that basis the Americans selected a site called Cerro Tololo, near the city of La Serena, about 500 km north of Santiago. ESO's interest led to the addition of André Muller to Stock's team, which then started to search for a site for ESO. In 1963 a delegation of ESO together with some Americans visited a very similar location as Cerro Tololo, called Cerro Morado. Oort was part of that expedition (see Fig. 10.8). In the fall of 1963, ESO decided to go to Chile because of the superior climatic conditions there for astronomical observations. Now that the convention had been ratified, Otto Heckman was appointed the first Director General of ESO and he started negotiations with Chile about the installation of a European observatory on Chilean territory. This led to the selection of the mountain La Silla, about 80 km north of Cerro Tololo, in May 1964. Not much later, the Carnegie Institution decided to build a southern observatory on Las Campanas, another 30 km north of La Silla.

Fig. 10.8 Oort on horseback in Chile in 1963, visiting a possible location for the ESO observatory, Cerro Morado, not far from where the Americans wanted to build an observatory, Cerro Tololo. Eventually ESO did not go to this place. This picture is part of a larger one that shows the whole company. Credit: ESO/F. K. Edmondson [111]

Oort and Baade are rightly seen as the founding fathers of the European Southern Observatory. The inauguration of La Silla took place on March 25, 1969. First a 1.5 m and a 1 m Schmidt telescope were delivered, but next to that smaller national telescopes were placed as well, of which ESO astronomers then could get a part of the observation time. The 1 m Schmidt telescope came into operation in 1971 and André Muller has done an enormous amount of work for it. Under the second Director General Adriaan Blaauw, the flagship, the 3.6 m Telescope, was built and put into operation in 1976. ESO's success to this day is spectacular.

10.4 More Dynamics

The dynamics of our Galaxy remained a primary interest of Oort and he continued to conduct or supervise research in that area. First of all he did a study of the globular cluster Messier 3 (all globular clusters look alike, so M3 looks just as M13 in Fig. 10.9). Oort had obtained – before publication—new, very deep counts of stars that Allan Sandage had obtained from plates with the 200 in. Telescope of Palomar Observatory. Oort did this work together with Gijs van Herk. The latter had moved first to Washington to work at the United States Naval Observatory in 1958 and on to Copenhagen for some time, but returned to Leiden in 1962. The work with the meridian circle in Leiden, which was van Herk's specialization, had already been discontinued for some time. His part of the work on the globular cluster had been done before he left for Washington.

The reason that these objects have so little structure, lies in nearby encounters of the stars in them (no collisions, which are extremely rare), where stars disturb each other's orbits and exchange energy in the process. As a result, any structure is smoothed out. The time scale on which this takes place is called the relaxation time and Oort had estimated that this was such that the inner 8 parsec would be 'relaxed' (something like the inner half of Fig. 10.9). In the outskirts, the orbits of the stars are more elongated. The available energy is distributed evenly among the stars and heavy stars will therefore move slower and be concentrated more towards the center than the lighter ones.

Oort also returned to the question of the vertical force in the disk of our Galaxy. Here, too, there were new data from Nancy Roman's work on K-giants in the vicinity of the Sun. This made it possible to determine their spatial distribution and the random motions better than before. The work was done by an Australian student, Eric Richard Hill (1927–2016), who had been sent to Leiden by his boss, the radio astronomer Joseph Lade Pawsey

Fig. 10.9 The globular cluster Messier 13 (M13). From Wikipedia [112]

(1908–1962) from Sydney, to gain expertise in optical and classical astronomy for his group. Hill arrived in 1952 and planned to produce a PhD thesis. The work resulted in a new run of the vertical force with distance from the plane of the Milky Way. The answer that Hill found was unrealistic above 500 pc; the force dropped down sharply and that is not physically possible. But close to the plane it was realistic and a result worth publishing and writing a thesis about.

Eventually—in 1956—Hill left and went back to Australia without a doctor's degree. With great delay he published the result in the *Bulletin of the Astronomical Institutes of the Netherlands (BAN)* in 1960. Oort did the analysis again and made other assumptions about the incompleteness of the set of observations and on how to correct for that. From his analysis, which also differed in other aspects from Hill's, he found an entirely different result and published it in a subsequent article. It may seem as if Oort felt that Hill's work was below standards. Although Oort and Hill differed in outcome, and disagreed on details of the approach, the extensive correspondence between Oort and Pawsey reveals a different picture. The reason for Hill's departure seems

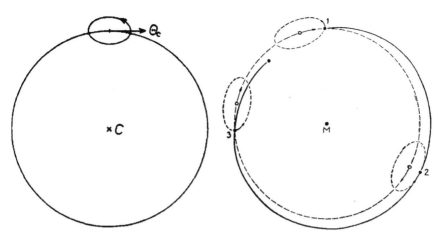

Fig. 10.10 Epicycles as a description of the orbits of stars with a small random velocity. On the left in a co-rotating system and on the right an inertial system. From publications by Oort in 1965 and 1943, respectively. From the Oort Archives

to have been nothing more than a severe case of homesickness. After Hill was back in Pawsey's group, both Oort and Pawsey urged Hill for a long time to publish his work, which eventually happened. If Oort had been convinced that Hill's work was substandard, he would never have allowed it to be published, let alone insisted on it.

Another study concerned the orbits of stars in the Milky Way Galaxy near the Sun. In general, stars do not go around the center exactly in pure circles, but have small deviations from pure rotation. If the deviation is small, you can describe it in the plane as a general circular motion around the center with superposed on that an oscillation in the radial and tangential directions. The dynamics turned out to prescribe that the frequency of this is the same in both directions in the plane and that the ratio of their magnitude depends on the Oort constants of differential rotation. Lindblad had described this as an epicyclic movement, analogous to the old representation of planetary orbits. Fig. 10.10 illustrates this. In the left-hand panel we move, as it were, along with the rotation in the large circle. The star describes an ellipse in such a system of coordinates. If we do not rotate along we get the right-hand panel. The orbit will then swing around the circle that represents the general rotation, but in general that orbit will not be closed. It can be closed if you rotate exactly the right way. For example, in the case that the center of the epicycle goes around the center once in the same time that the star goes around twice in its epicycle. This would result in a closed, oval orbit. This is called the inner Lindblad resonance.

In addition, stars will also have a vertical oscillation, which in practice will be independent of the epicycle oscillations in the plane. For the Sun, the average distance to the Galactic center is 8.2 kpc and the Sun (and we with her) goes around the center of the Milky Way in 240 million years. The dimensions of the epicycle are 0.48 by 0.34 kpc and the period therein is 170 million years. In the vertical direction the period is 80 million years and the maximum distance to the plane is 85 parsec. So the motions in and perpendicular to the Galactic plane are completely decoupled.[1]

The Galaxy is very flattened. For stars like the Sun, which come only a small distance from the plane, the vertical oscillation—as we have seen—is completely disconnected from that in the plane. The gravitational force towards the center does not change perceptibly if one moves a little bit out of the plane. This is described as the 'third integral'. An integral of motion is something that is constant for a star everywhere in its orbit, but that is not always something that has a physical meaning. There are six of such integrals of motion.[2] Two of them are known, namely the total energy and the angular momentum (the amount of rotation); the others generally have no physical interpretation. In the case as described, the total energy in the vertical direction (kinetic plus potential) is also conserved. This is then a third integral.

But the question now was what happens when the velocity increases and the star deviates further from the plane. Alexander (Alex) Ollongren (b. 1928) made a study of this with Henk van de Hulst at Oort's suggestion. Among others, he had Swedish roots and could arrange therefore that he could use one of the then fastest and most powerful electronic computers in the world—BESK for Binär Elektronisk SekvensKalkylator—in Stockholm (the ARMAC, Automatische Rekenmachine MAthematisch Centrum, in Amsterdam, which he also used, was not fast enough for a full calculation). Ollongren's work was funded by ZWO. One result can be seen in Fig. 10.11, where an orbit in a plane is shown perpendicular to the disk and rotating along in the tangential direction (like Oort, he used the symbol 'curly pi' ϖ for the radial direction R). It resembles two oscillations in a distorted system of coordinates that are independent. The coordinates are called confocal ellipses and consist of ellipses and hyperbolas instead of straight lines. So there is indeed a third integral that can be expressed in this coordinate system.

---

[1]The Sun is currently 13 parsec from the plane of the Milky Way. I have pointed out elsewhere that if the biological evolution had proceeded 0.5% faster or slower since the formation of the Earth, we would have been outside the dust layer and then we would have had an unobstructed view of the center of our Galaxy.

[2]Mathematically, the equations of motion are of the form force equals mass times acceleration; the latter is a second derivative of time and since there are three coordinates, we have to integrate six times to solve them. Then there are six constants of integration.

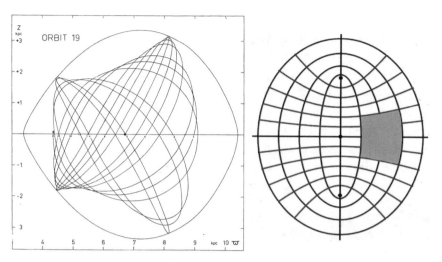

Fig. 10.11 On the left Ollongren's 'Orbit 19' in the R-z plane. Here the old-fashioned 'curly pi' ϖ is used instead of R. On the right: confocal ellipsoidal coordinates. From the Oort Archives, and adapted from Wikipedia Commons [113]

There also finally came a conclusion to the work on photographic surface photometry, which Oort had started at Perkins Observatory and had continued with the observing session at Mount Wilson in 1939 and the photoelectric calibration observations with Hiltner at McDonald Observatory in 1947. The work was done by Cornelis Johannes (Kees) van Houten (1920–2002), who reduced the data and also had a further observing session at McDonald Observatory. This resulted in a PhD thesis in 1961 with detailed brightness distributions in twenty galaxies, most of which were seen edge-on. This was an enormous achievement, but Oort did not do any more follow-up studies with it. For years it has remained the largest, homogeneous collection of surface brightness distributions in galaxies. An important system in van Houten's thesis was NGC 3115, which Oort had discussed at the McDonald inauguration in 1939. But insufficiently reliable velocity information prevented a conclusive analysis, despite van Houten's measurements.

Kees van Houten also did not continue this work. He had moved to the Sternwarte Heidelberg-Königstuhl, where he and his German-born wife Ingrid Groeneveld (1921–2015) worked together on asteroids. They had met each other when they both worked at Yerkes with Gerard Kuiper and were married in 1955. When Kees returned to Leiden, Ingrid accompanied him and after Kees' PhD defense they moved to Heidelberg. Eventually they would return to Leiden, where they worked the rest of their careers on asteroids.

The epicyclic description has an interesting consequence. In a coordinate system that rotates at such an angular velocity that in it the star goes around the center exactly once in the time that it goes through its epicycle twice, a closed orbit results. Now it turns out that the angular velocity required to produce such an inner Lindblad resonance in the Milky Way Galaxy is very closely the same for a large range of distances from the center. So we then have the situation, that we can define a rotating coordinate system in which over a large range of galactocentric distances stars move in more or less closed, somewhat elongated orbits; if then there is a pattern in those orbits, then that pattern

Fig. 10.12 Persons who played a role in radio astronomy in the Netherlands in the 1960s. Top row from left to right: Dini Ondei, for many years Oort's secretary, King Kwee, Oort, Pieter Oosterhoff, Gail (Bieger–) Smith, and Gart Westerhout, between Kwee and Oort Louise (Gelato–)Volders, and between Oosterhoff and Smith, Ben Hooghoudt. Middle row: Jan Högbom, Whitney Shane, Jaap Tinbergen, Alex Ollongren, Ernst Raimond, and Wim Brouw. Bottom row: Jet Merkelijn, Luc Braes, Hugo van Woerden, Elly Berkhuijsen, Butler Burton (behind him Kees van Houten), and Wim Rougoor. Top and middle row from a group photograph of the Sterrewacht staff in 1960. Bottom from the group photograph of participants at a symposium in Noordwijk in 1966. From the Oort Archives

could continue to exist. This was found by Bertil Lindblad and his son Per Olof (b. 1927). We will come across this again when I discuss spiral structure further.

10.5   Research in Dwingeloo

Part of the research with the Dwingeloo Radio Telescope until the early 1970s was performed in the context of PhD dissertations. Some had Henk van de Hulst as supervisor. Gart Westerhout's thesis concerned the radio continuum radiation from the Milky Way. The continuum radiation over wide areas of the sky and its polarization were mapped by Michael Moore (Mike) Davis (b. 1938) from the United States, Elisabeth Mabel (Elly) Berkhuijsen (b. 1937), Willem Nicolaas Brouw (b. 1940), Jaap Tinbergen and Titus Adrianus Thomas Spoelstra (1945–2010). This yielded, for example, information about the magnetic field in the Galaxy, which was of great interest to Henk van de Hulst. Brouw and Tinbergen wrote theses on other subjects. Berkhuijsen was supervised by Oort, Spoelstra with Oort as one of two supervisors, the others with van de Hulst. Many more details about the Dwingeloo Radio Telescope and its research can be found in [114, 115]. Pictures of persons involved in early Dwingeloo studies have been collected in Fig. 10.12.

Oort's interest was mainly in observations of the 21-cm line. The first studies using this line concerned special areas in the Milky Way, resulting in the theses of two students, who had been involved in radio astronomy in Kootwijk from an early time. The first was Ernst Raimond (1932–2010), son of Jean Jacques Raimond, who had Oort as thesis supervisor, and the second one was Hugo van Woerden (b. 1926–2020), who defended his dissertation in Groningen with Adriaan Blaauw. Oort's greatest interest was in three subjects—the Galactic Center, spiral structure and high-velocity clouds. I will deal with these in this order.

Already in Kootwijk it was found that there was a whole range of velocities in the hydrogen in the direction of the Galactic Center, while it was expected to be very limited. After all, the rotation along the line to the center is perpendicular to it and radial velocities are expected to be zero. It turned out that there were two 'arms' that moved away from the center. On our side it is the so-called 3-kpc arm (because of the estimated distance from the center), which moves towards us at 53 km/s. The core of our Galaxy contains a strong radio source (Sagittarius A) and the hydrogen was therefore seen in absorption at the corresponding velocity. On the other side, there is an arm at 135 km/s, which was not seen in absorption and therefore had to lie behind the center. This discovery in Kootwijk by van de Hulst, Raimond and van Woerden were followed up with

Dwingeloo observations in the PhD thesis of Gerrit Willem (Wim) Rougoor (1931–1967) in 1964 with Oort as supervisor. Oort speculated that this was the result of an explosion in the core of the Galaxy, in which clouds of gas had been ejected that were slowed down by the gas in the disk.

When I—Pieter Corijnus van der Kruit (b. 1944)—had obtained my candidate degree (bachelor) in 1966, Wim Rougoor just started a research project in Dwingeloo observing in an area around the Galactic center, but outside the plane of the Milky Way and I was working on it for a possible master's thesis. Unfortunately, Rougoor died a year later of leukemia; after that, I continued the work by myself under the supervision of Oort. Soon we found that gas clouds were moving outside the plane of the Galaxy, one on one side of the center away from us (which was already known), but symmetrically opposite on the other side also one coming towards us. This seemed to be a convincing indication of gas cloud expulsions due to activity in the nucleus. Such activity was also found in so-called Seyfert galaxies, named after Carl Keenan Seyfert (1911–1960), spiral galaxies in which bright nuclei showed large internal motions of the gas. Eventually, this became my PhD thesis, which I successfully defended with Oort as my supervisor in 1971 (Fig. 10.13). It also contained calculations of how gas clouds expelled from the nucleus at an angle to the plane could give rise to the expanding arms after they had fallen back into the plane a few kpc from the center and been slowed down. Nowadays, the usually accepted explanation is that gas moves in oval orbits as a result of the asymmetric gravitational field of a central bar in the Galaxy.

Spiral structure was the second subject of study, mainly based on Dwingeloo observations in the PhD theses of William Butler Burton (b. 1940) and William Whitney Shane (b. 1928), who were American students that came to Leiden to obtain their degrees. Shane is the son of the former director of Lick Observatory, Donald Shane. There was a further American student, Gail Patricia (Bieger–)Smith (b. 1939), who left astronomy after her marriage to a Dutchman. The work is within the framework of the density wave theory, which I will first explain.

This theory focuses on the long term maintenance of the spiral structure, at least for those cases where a pattern extends over the whole system and arms can be followed from the inside out to the outskirts. This is seen as a wave pattern that rotates more slowly, but without differential rotation. It then retains its shape. This is probably related to the property noted above, that the angular velocity of a rotating coordinate system, in which an inner Lindblad resonance gives closed orbits, is constant over most of the Galactic disk. Stars and gas move in relation to the spiral wave pattern, overtaking the arms. These are then slightly accelerated by gravity due to the greater density in the arms in

Fig. 10.13 Oort and myself at the reception at the Sterrewacht after the defense of my PhD thesis in October 1971. New doctors were allowed after the defense in the Academy Building to walk through the Hortus Botanicus (the oldest botanical garden in the Netherlands, dating back to 1590) with their guests to the Sterrewacht. The thesis defense takes place in dress suit and the professors wear academic gowns. Oort must have taken his off in the Academy Building before he joined that walk. The tube at the bottom right contains the doctor's diploma. From the author's collection

front, but slowed down again when they have passed through. This gives a new concentration, so that the density wave is maintained, as it were. But because of those small accelerations there is a pattern of streamings, which corresponds with that of the spiral structure. In systems without large scale spiral structure, larger areas of star formation are stretched into small pieces of spiral arcs by differential rotation.

Now it also became clear, that the spiral pattern of Fig. 9.6 is in fact incorrect! The reason lies in the assumption that a peak in the observed profile (brightness of the 21-cm line with wavelength or radial velocity) must also be a peak in the density along the line of sight. This is not necessarily the case when the velocity does not vary uniformly along this line. If somewhere the change is fast, the signal will be distributed over more wavelengths than when the variation is small. Now, the effects of this, especially in the presence of streaming motions due to density waves, turned out to be substantial. Burton developed a computer program (meanwhile, Leiden University had an IBM computer at the end of the 1960s, which we PhD students used extensively in our research) that could simulate the effects of density waves and calculate (simulate) the corresponding 21-cm observations. This can then be compared to actual Dwingeloo observations. Burton found an excellent explanation of all the large-scale patterns in the observations with a density wave model of our Milky Way System.

Shane also found such motions in one arm closer to the center (but outside the 3-kpc arm) and showed that there could also be possible explanations of the expansions, if gas moved in oval orbits oriented in such a way that along the line to the center the gas in it moved away from the center and in the perpendicular direction towards it (which is not in our line of sight so that we cannot observe it). The picture that emerged from this and from more detailed work later on is that our Galaxy is a barred spiral with a large-scale spiral pattern, as for example in the UGC 12158 system in Fig. 10.14. There is not yet a reasonable explanation for the expanding clouds outside the plane in the central parts of the Galaxy in the context of this theory.

What interested Oort greatly were the high-velocity clouds. Spitzer had suggested that the Galaxy is surrounded by a gaseous corona, which is hot but in which neutral condensations might occur. Such clouds were eventually discovered around 1960 in Dwingeloo when the receivers had become sensitive enough. In Leiden this work was mainly done by Ernst Raimond and student Adrianus Nicolaas Maria (Aad) Hulsbosch (b. 1941). Remarkable was that these high-velocity clouds were concentrated in the sky in an area of about 90° in Galactic longitude by 30° in latitude, but even more curiously, they all had negative velocities, so were moving towards us. The largest example is called HVC Complex A around Galactic longitude 155° (i.e., seen slightly outward from the Sun) and latitude 40°. It is coming towards us at 165 km/s and covers a few degrees in the sky. The distance was completely unknown; if this were, for example, at 1 kpc, the mass would be about 2500 solar masses. More recent studies of stars in the direction of Complex A by Hugo van Woerden and collaborators indicate that it must be behind those stars and therefore at least a few kpc away from us. Material with lower velocities is also common, again mainly at negative velocities, as was shown by work in Groningen led by Adriaan Blaauw and Hugo van Woerden in the 1960s.

Oort wrote an extensive article in 1966, in which he discussed various possible explanations for the high-velocity clouds. Nearby very old supernova remnants, satellite systems to the Galaxy in which star formation had not taken place, gas ejected by the Galactic center, and a few others were excluded.

Fig. 10.14 The barred spiral system UGC 12158, photographed with Hubble Space Telescope. The central bar and the spiral structure may be similar to those in our own Galaxy. Credit: ESA/Hubble and NASA [116]

He concluded that the most likely explanation had to be that this gas fell from the Universe into the Galaxy. Our Galaxy would then move through an intergalactic medium, and the direction in which this happened would then result in the concentration around Complex A. Oort asked me to do calculations on that and the result was encouraging; there would even be an area in the sky where high-velocity clouds would not be allowed to occur, because clouds there would have to have an orbit that would have taken them through the disk of the Galaxy first. This matched the observations. Although I drafted an article and Oort edited it extensively, it was never published.

In the meantime it has become clear that there are three different kinds of high-velocity clouds. A part is the result of gas expulsion from the disk during supernova explosions, which is hot and ionized and thus invisible, but cools and recombines when it falls back to the disk. This is called the Galactic Fountain. Another part would be indeed in-falling intergalactic gas and yet another part results from interaction of the Galaxy with the Magellanic Clouds and these objects among themselves. Aad Hulsbosch did most of the Dwingeloo work on these clouds; his dissertation took more than ten years to complete, so he only got his PhD with Oort in 1973.

There are exceptions to the negative velocities. A good example is the Smith Cloud, discovered by Gail Smith and Whitney Shane. This is moving at 90 km/s *away* from us. However, it has also recently become clear that this is a cloud of gas falling in from the intergalactic medium, but unrelated to the primary flow around Complex A. It falls as seen from our positions into the disk on the other side of the center. Gail Bieger-Smith later attracted some attention in the media. She would have remarked in an interview [117][3] that Oort once told her that there was no room for women with children in astronomy and that she therefore had to leave astronomy. This turned out to be incorrect, as she told me during a telephone interview I conducted with her in 2017; she had left astronomy, on her own initiative, when she was pregnant and Oort had in fact given her support to finish her work. Nothing points to a prejudice by Oort against women, nor mothers, in astronomy. However, it is true that Oort did expect his staff (of both sexes) to be fully dedicated to the profession.

The discovery of the in-fall of intergalactic gas, which would not have been 'polluted' by the first stars with elements heavier than helium, made Oort think about the formation of galaxies. Oort estimated that the inflow into the disk at this time would be of the order of 3×10^{17} H-atoms per cm^2 per million years, and that corresponds to a few thousand solar masses per square pc per million years, or a 10% or so of the mass of the Milky Way over its total age. So gas inflow must have been a significant factor in the evolution.

An important aspect of the formation of galaxies is the origin of the angular momentum, the rotation of galaxies. How do galaxies obtain their rotation? There was a suggestion that this was due to tidal interaction in the early Universe when proto-galaxies were closer together and irregular in shape. As a result, the attraction exerted is not the same everywhere and rotation results. But the problem was that with the current number of galaxies and their estimated mass this effect would have been far too small. Oort proposed that there must have been many times more proto-systems than the ones we see now. Then the effect of that tidal effect would have been much greater. But then there must be still more matter now, and Oort suggested, that it would have to be in the form of hot gas, which could only be seen at X-ray wavelengths. Oort assumed that the moment at which concentrations became independent of the expansion of the Universe and detached themselves from the rest, and tidal interaction and the formation of galaxies would take place, would be quite late (perhaps only a quarter of today's age).

---

[3]For more details see [2].

Oort made the following estimate. Take a cloud of gas, which can become a galaxy and has rotation. In the direction perpendicular to the rotation, there is only gravity inward and contraction will eventually occur. In the plane of the rotation there is centrifugal force and contraction will not always occur. Now can take four perpendicular directions and along each of those directions the motion can be systematically inward, but also equally probably outward. Only if in all four directions these are inward, a galaxy can form, and because there is no reason why those directions would not be randomly distributed, the chance is one in sixteen. After al, sixteen different combinations are possible, of which only one works. So, there would have to be fifteen times more mass in the Universe than present in galaxies.

This theory of Oort was soon overtaken by new developments. Especially in the Netherlands with the Westerbork Synthesis Radio Telescope, but also elsewhere, it was found in the 1970s that the gas in spiral galaxies often stretches out far beyond the optical image and rotates in these outskirts just as fast as in the inner parts. So that rotational curves are 'flat'. This indicates that galaxies are much larger and contain huge amounts of matter that cannot be observed other than by gravity, the so-called dark matter halos. As a result, galaxies are at least ten times heavier than would be deduced from starlight and the origin of angular momentum could still result from tidal interaction in the early Universe without having to invoke the existence of condensations that failed to become galaxies with stars. So Oort was correct in inferring from the rotation of galaxies that there is more matter than meets the eye in galaxies, but while he supposed the solution was that it was concentrated in more (invisible) objects it turned out that galaxies contain much more mass.

Fig. 11.0 Oort during two of the ceremonies in which he was awarded honorary doctorates. On the left at Leuven in 1955 and on the right at Cambridge (U.K.) in 1980. From the Oort Archives

11

Westerbork and Retirement

You can always amend a big plan, but you can never expand on a little one.
I don't believe in little plans, I believe in plans big enough
to meet a situation which cannot possibly be foreseen now.
Harry S. Truman (1884–1972)

Near Westerbork a gigantic Synthesis Radio Telescope was being built,
consisting of twelve dishes, [...]
It was going to be the biggest instrument in the world
and he expected a lot of it.
Harry Mulisch

11.1 Personal Matters

An important change in the life of the Oorts came in 1966, when their younger son Abraham (Bram) Oort and his wife Bineke Pel and children emigrated to the United States. At their wedding in 1961, Oort had addressed them as an astrologer (see Fig. 11.1); he had collected their character traits, as they matched their astrological zodiacal signs and the positions of the planets at birth, and predicted that they suited each other very well, because both liked good food and drink. Bram worked at the Geophysical Fluid Dynamics Laboratory, first in Washington D.C. Two years later they moved to Princeton and that gave Oort a good opportunity to visit them annually, since he could always arrange to work at the Institute of Advanced Study. Oort and Mrs. Oort kept that up until 1985 when traveling became too difficult due to advancing ages.

© The Editor(s) (if applicable) and The Author(s), under exclusive license
to Springer Nature Switzerland AG 2021
P. C. van der Kruit, *Master of Galactic Astronomy: A Biography of Jan Hendrik Oort*,
Springer Biographies, https://doi.org/10.1007/978-3-030-55548-1_11

Fig. 11.1 Oort as an astrologer at the wedding of his younger son Abraham to Bineke Pel in 1961. From the collection of Abraham H. Oort

In the 1960s the number of grandchildren grew rapidly. Figure 11.2 stems from shortly before the emigration of Bram and his family. To the six boys and one girl two more girls and one boy would be added. Of the generation of the parents of Oort and his wife, only Mrs. Oort's mother was still alive at the time.

Oort's mother, Ruth Faber, had died in 1957. She was cremated at the Westerveld cemetery at Driehuis near Velsen. This exists since 1888 as the first cemetery for all religions and included a crematorium from 1914 onward. Cremations were at first formally illegal, but were tolerated until 1955, when it became legal. There are no documents of the death of Oort's mother, only an advertisement in the local newspaper Leids Dagblad. Mrs. Oorts mother, Helèné van de Water, (grandma Lène in the family), died in 1974 at the age of 93.

Here is part of a note by Bram Oort about his parents:

Fig. 11.2 Four generations in the Oort-family in 1965 or 1966. In the back in the middle Oorts mother-in-law Heleen Graadt van Roggen-van de Water, to the right daughter Marijke and son Coen, in the front on the left Bram. In the back at the left their partners Bineke Pel, Marianne Lissy and Tom de Smidt. The eldest grandson Marc Oort, who would later write an astronomical dissertation, sits in the front in the middle. From the collection of Abraham H.Oort.

My parents loved flowers; there was always a bouquet on the coffee table and my father's desk. In the summer we had the flowers from the garden; in the other seasons my parents went by bike to the Saturday market to buy them. My mother did most of the planting and the weeding. There was an abundance of red and pink roses against the lattice outside my father's official office. He was in the habit of clearing dead or nearly dead flowers in the bouquets, not only in our home but also in those of friends (a source of entertainment for us). Later, when they lived in their first own house in the Kennedylaan in Oegstgeest, they maintained a beautiful garden. They both enjoyed working in the garden.

Another piece of Bram Oort about the Oort family:

Coen and I had little in common and we grew further apart after our emigration. He was extraverted, noisy and a talker, while I was introverted, calm and quiet. He loved to live in a complicated, luxurious way, while I loved a simple life. When I was little, he used to tease me with my games when I played with my dolls that I was a teacher or a minister. On the other hand, I admired him because

Fig. 11.3 The Oorts, probably in 1956. They loved to read books to each other in the evening. This may be an example of such an activity. From the Oort Archives

he could play the cello so well (I loved listening to him). He also helped me with translations of Latin texts from Caesar's 'De Bello Gallico. He was a good teacher.

I also enjoyed Marijke's singing and piano playing at the Sterrewacht when I went to high school, and later when we visited the Netherlands. With a good friend she sang 'Sound the Trumpet' by Purcell at our wedding in Den Haag in 1961. Marijke felt it was terrible that we emigrated to the United States (she always would have liked to have a sister like in Bineke). The last time we saw her in good shape was on her 75th birthday in 2006. Two years later she died unexpectedly of a fast-growing abdominal cancer. I was with her during her last days (she recognized me but could not talk anymore). It was sad that the last of my family members died.

For many years, the Oorts organized parties at the Sterrewacht for the staff in order to promote relations and contacts between the academics and the other employees (such as calculators and instrument builders). Mrs. Oort usually took the initiative and Oort came up with a theme. One was held in the summer and one in the winter. In the 1960s this apparently stopped, because I did not experience any of it myself. Especially when the children grew up the Oorts liked to read books to each other (Fig 11.3).

11.2 Honors, Golden Quill and Vetlesen Prize

Appendix B.1 lists the honors received by Oort. Among them are 22 member-ships of Academies of Science and other scientific societies in sixteen countries, and 10 honorary doctorates at foreign universities. Important medals, which we have already seen, are the Bruce Medal of the Astronomical Society of the Pacific in the U.S., and the Gold Medal of the Royal Astronomical Society in the U.K. I will focus below on two awards from the 1960s, the 'Golden Quill' of the Royal Dutch Publishers Association, and the very prestigious Vetlesen Prize of Columbia University. Oort received two Royal distinctions; as we saw he became Knight in the Order of the Dutch Lion in 1956 at the inaugura-tion of the Dwingeloo Radio Telescope; when he retired in 1970 he was made Commander in the Order of Orange-Nassau. Further on in this book I will come back to two prizes that he received in the 1980s, the Balzan Prize and the Kyoto Prize.

Appendix B.2 contains Oorts academic genealogy. An extended version with references and links can be found on my homepage (because as a student of Oort it is also my genealogy) [118]. In principle, these involve academic dissertations, with which a PhD degree has been obtained with the previous person as supervisor. But that is not always possible, especially longer ago. But then there is often a professor-student relationship. The genealogy ends with Philipp Müller (1585–1659), professor of mathematics at the University of Leipzig from 1616 onward. Müller had a keen interest in astronomy. He corresponded extensively with Johannes Kepler (1571–1630) and was one of the first to adopt Kepler's laws of planetary orbits. Jacob Bartsch (c.1600–1633), a pupil of Philip Müller, is known to have made a star chart based on Müller's data and later became Kepler's assistant and son-in-law. Probably Oort was unfamiliar with this. But it would undoubtedly have given him satisfaction (as it does me as a student of Oort) that it goes back to someone so close to Kepler (Fig. 11.4).

In 1960 Oort was awarded the 'Gouden Ganzeveer' (Golden Quill). It was first awarded in 1955 by the Royal Dutch Publishers Association to a person (or organization) who had been of special importance for the written and printed word in the Dutch language. The first recipient had been the KNAW, the second a historian in 1957, and Oort was the third laureate (Fig. 11.4). In the meantime, this distinction is awarded annually by a special foundation, gen-erally to successful authors of novels and other books. In his motivation, Oort was not only praised for his astronomical research and administrative work, but especially for his popularizing articles, such as in *Hemel & Dampkring* and the *Nederlands Tijdschrift voor Natuurkunde* (Netherlands Physics Magazine of the

Fig. 11.4 Oort receiving the 'Golden Quill' on October 13, 1960 from the Minister of Education, Arts & Sciences Jozef Cals. From the Oort Archives.

professional society of physicists). Indeed, Oort regularly wrote for these magazines. Besides in writing, for which this prize is awarded, Oort also gave many lectures and popular presentations. The Oort Archives contain about eighty sets of notes and comments for such lectures, dated between the mid-1920s and mid-1980s, and there must have been many more.

In the Netherlands every year a week or ten days is designated Book Week, dedicated to literature. Book shops have special deals, there is a theme and an author is asked to write a short novel or other book, often dedicated to the theme and this 'book week gift' is given free to anyone spending more than a certain amount in buying books. The Book Week has existed in the Netherlands since 1932. Since 1947 it is preceded by the 'Book Ball', held in the Concertgebouw in Amsterdam. Oort and Mrs. Oort were invited (perhaps several times). In Fig. 11.5 they are in conversation with Queen Juliana. The copy of this photograph in the Archives has a note on the backside saying 1960; however, the Book Week is always held in March, and Oort received the Golden Quill in October 1960. If Oort's presence at the ball is related to the award, it must have been 1961.

The Vetlesen Prize has been awarded every two years since 1960 to individuals who have made a special contribution to our understanding of the Earth, its history and its relationship to the Universe. It was established by the

Fig. 11.5 Oort and Mrs. Oort talking to Queen Juliana at the book ball. In the background Harry Mulisch, the later author of *The discovery of Heaven* (see quotes at the beginnings of chapters 9, 10 and 11). From the Oort Archives

Norwegian-American shipbuilder Georg Unger Vetlesen (1889–1955). The prize went except for a few rare exceptions to geophysicists and geologists (such as national meteorological service KNMI director Felix Vening Meinesz) and has a status comparable to that of the Nobel Prize. The fact that Oort's student Lodewijk Woltjer worked at Columbia will have contributed, but despite this, the prize was later awarded almost exclusively to Earth scientists and not to astronomers. Apart from being extremely honorable (and accompanied by a gold medal), it is also a large prize in terms of the amount of money involved, 250,000$ (a value now of about 1.3 million €).

The award ceremony took place on October 26, 1966, followed by a symposium on *Galaxies and the Universe*. At the award ceremony Oort's good friend and colleague, Bengt Strömgren presented a laudation in which he praised Oort for his work on the kinematics and dynamics of the Galaxy and the interstellar medium, but also for his great dedication to the International Astronomical Union.

At the symposium Oort talked about the work with the Kootwijk and Dwingeloo radio telescopes and in particular what interested him most - the spiral structure, the Galactic Center and the high-velocity clouds. The other speakers presented talks on subjects of his special interests of the time. Bruno Benedetto Rossi (1905–1993) of the Massachusetts Institute of Technology spoke about the new possibilities of the emerging X-ray astronomy. Chia-Chiao Lin (1916–2013), also from MIT, presented the density wave theory of spiral structure, which he had developed not long before that with Frank

Hsia-San Shu (b. 1943). Another topic that fascinated Oort, the 'timescale of creation', the determinations of and comparison between the ages of the oldest stars in globular clusters and the disk of the Galaxy, the chemical elements from radioactive isotopes in meteorites, and of the Universe itself from its expansion, was discussed by Allan Sandage.

Oort never bragged about his prizes and honors. Back in Leiden after the Vetlesen Prize ceremony he organized a big party at the Sterrewacht, emphasizing that this was the result of a joint enterprise of many and a recognition of their work and dedication.

11.3   The Westerbork Synthesis Radio Telescope

Even before the Dwingeloo Radio Telescope was ready for observing, Oort thought about a larger radio telescope. The problem with radio astronomy is its resolving power. You cannot see as sharp as in optical light. The sensitivity of a the telescope is largest in the direction it is pointing. If we set that equal to one, then at an angle away from this the sensitivity will be smaller. The angle between the positions where on either side of the center it is one half is called the beam. It arises because all the waves reflected by the various parts of the 'mirror' come together in the focal point and Interfere, no < there to form an image. If two of these waves are 'in phase', they reinforce one another; if they are completely 'out of phase', they extinguish each other. The size of the beam (expressed as an angle) is inversely proportional to the diameter of the dish and proportional to the wavelength. For Dwingeloo this gives (at 25 m diameter of the dish) at 21 cm wavelength a beam with a diameter of half a degree, about the size of the Moon in the sky.

The technique of interferometry now was to replace the dish with a number of smaller elements and then bring the signals together electronically. In Fig. 11.6 we see a two-element interferometer consisting of two element radio telescopes or antennas; the signals are brought together through cables so that they 'correlate', making sure that they arrive at the receiver at the same time (by sending the signals through cables of unequal length; nowadays this is done electronically). In the direction in which the elements are aimed, these signals then reinforce each other (the wavelength difference d_1 is then compensated), but not in a slightly different direction (d_2). For example, if the difference between d_1 and d_2 is half a wavelength, they extinguish each other. This simulates the process of imaging in the focus of a single telescope in the direction parallel in the sky to D.

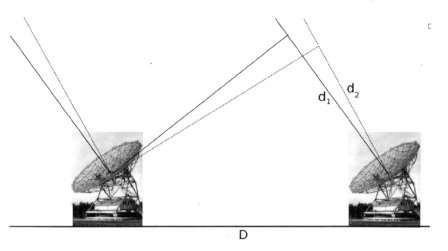

Fig. 11.6 Signals from two telescopes at a distance D from each other correspond to a difference in the distance traveled. For two directions in the sky this is d_1 and d_2. Figure by the author

This technique of interferometry was used in many places, notably in Australia and the U.K., such as the later Nobel Prize winners Martin Ryle (1918–1984) and Antony Hewish (b. 1924). They and Alfred Charles Bernard Lovell (1913–2012) were radio engineers who had worked on radar techniques during the war. Lovell went to Manchester and built a 250-foot (72 m) radio telescope at the Jodrell Bank site, making Dwingeloo the largest radio telescope in the world for only a few months. Ryle and Hewish went to Cambridge and first built a rectangular interferometer of 580 (east-west) by 52 m (north-south) with antennas in the corners. In 1955 they found this way 1936 radio sources all over the northern sky.

These were uniformly distributed across the sky and therefore had to be outside the Milky Way Galaxy. If intrinsically they were all equally strong on average, you would expect that if you go twice as far away there would be four times as many, which would be seen as four times fainter. This gives a predicted increase in the number of sources with decreasing observed brightness. If they are not all equally strong, but have the same distribution at different distances, then that prediction is the same. But Ryle and his (originally German) student Peter August Georg Scheuer (1930–2001) found a stronger increase at weaker levels than this prediction.

The expansion of the Universe indicates that there has been a beginning to the Universe, and for some this was an unsatisfactory situation. They argued that the Universe should not change, and thus matter had to be created constantly. Among them was Fred Hoyle. In fact, this was the reason why he had done so much work on the production of chemical elements in stars, because the assumption was that that new matter was in the form of hydrogen. Ryle and collaborators concluded that their observations showed that there had to be cosmic evolution in the population of radio sources (there were many more at larger and larger distances and thus further back in time) and this ruled out this 'steady-state theory'. This became a controversy that caused a lot of debate in Cambridge, which became more and more unpleasant. Oort was fascinated by Ryle's result and wrote him in a letter that 'this was the most fascinating discovery I have ever seen'.

It was this kind of work that Oort had in mind for a new radio telescope, and decided it had to be based on interferometry. At first he thought of something like 'Mills Cross', a telescope in the form of a cross that Bernard Yarnton Mills (1920–2011) had built at Fleurs near Sydney in Australia. In 1951 Oort appointed a young American radio engineer, Charles Louis Seeger (1912–2002), and in 1958 he asked him to design a radio telescope for the Netherlands following the concept of Mills' Cross, which consisted of two arms with a curved, reflective screen behind a row of dipoles. However, it soon became clear that it would be too expensive for the Netherlands and Oort tried to interest Belgium in working together. This is how the Benelux Cross Antenna Project (BCAP) came into being. Luxembourg was not involved, as the name Benelux would suggest. Oort secured more foreign employees, especially Wilbur Norman (Chris) Christiansen (1913–2007) from Sydney, Jan Arvid Högbom (b. 1929), a Swede who was a student of Ryle, and the American William Clarence (Bill) Erickson (1930- 2015). Seeger and these three all worked in the Netherlands for shorter or longer periods of time. But a huge effort also came from Dutchmen, especially Ernst Raimond and Ben Hooghoudt, both of whom had been involved with Dwingeloo. The concept was developed into a (second) design that is shown in Fig. 11.7.

The BCAP story covers a long period of time with some ups and many downs, which I will not recite here in detail (but see [119] and [9]). It took Oort a lot of effort. In the end, Belgium withdrew. But meanwhile Ryle in Cambridge had developed the technique of Earth-rotation aperture synthesis. This meant that he connected a set of antennas on an east-west line and then followed the sky for twelve hours. This meant that the combinations of elements included not only various distances between element antennas, but orientations as well. This allowed the simulation of image formation as in the focus of a single mirror. Ryle had tested this first with an east-west arm of an existing radio telescope built according to the principle of Mills' Cross. He now divided this arm into elements which he made to work as various interferometers as in Fig. 11.6. Because he had to follow the position

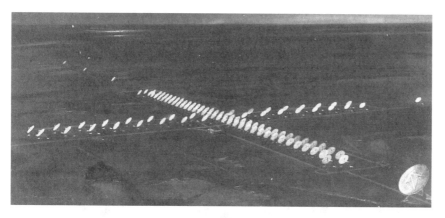

Fig. 11.7 The 'second design' of the Benelux Cross Antenna, dating from 1961. The arms were about 1.5 km long; the cross consisted of about a hundred antennas with a diameter of 30 m. From the Archives of the Sterrewacht Leiden

in the sky for twelve hours and the elements could not do so, he directed the elements towards the North Pole, which after all stands still (keeps the same direction, but rotates during the day). The result was a map with a radius of 4° around the North Pole, which showed 87 radio sources in that relatively small area. This left an enormously deep impression on Oort. Ryle developed several further interferometers, which culminated in 1964 in the 'One-Mile Telescope', consisting of two 60 foot (about 18 m) antennas and a third one on a rail-track, all east-west on a one mile long line. Observations of the same part of the sky are then repeated several times with the movable element in different positions. He was awarded half the Nobel Prize for physics in 1974 for this technique (Hewish was awarded the other half for the discovery of pulsars). As a result, the third design in the BCAP Cross antenna project became an east-west line of antennas.

That third design, drawn up in 1963, included 28 antennas on a 1.5 km east-west line and a further 6 antennas 1.5 km further along the line. The original idea had been to use a longer wavelength for source counts, but 21 cm was now again the primary wavelength (other wavelengths remained possible of course). The beam would then be about 10 arcsec east-west (north-south it depends on what the declination of the source is, it is also 10 arcsec in the pole and increases the closer the source is to the equator). The Netherlands (i.e. via ZWO) soon committed itself for half of the estimated costs of 30 million guilders (84 million € today in purchasing power). But Belgium remained hesitant. The scientific priorities there were different, and the Belgians wanted more international partners to reduce costs and also wanted guarantees for involvement of Belgian industry. In addition, Belgium had built

Fig. 11.8 The Westerbork Synthesis Radio Telescope, photographed from the west, not long after its inauguration in 1970. The building on the left is a large hall where the element telescopes were built. Hidden halfway along the the line on the right in the forest is the control center. The telescopes are 25 m in diameter, ten fixed in a row and at the far end two more on a rail-track. The total length is 1,5 km. Later two more antennas were placed on a rail-track near the top of the picture, making the total length 3 km. At the top left the remains of the Westerbork concentration camp, at that time still inhabited by a part of the Moluccan community. From the Archives of the Sterrewacht Leiden

a research station in Antarctica as part of the international geophysical year (1957–58) for which the Netherlands participated in an expedition from 1963 onward. The Netherlands withdrew from this expedition after a few years, so it had to be discontinued and that did not help to reach a positive decision in Belgium either. In 1967 Belgium formally withdrew from BCAP, but the

Fig. 11.9 The element antennas of the Westerbork Synthesis Radio Telescope seen from the east with in the foreground the two movable antennas on a rail-track. Provided by ASTRON, Dwingeloo

prospects of actual construction by the two countries together had already become minimal long before that.

As a result, a new design had to be made for half the price. This design started out with nine 25 meter antennas (like the radio telescope in Dwingeloo) on a 1.2 km long east-west line and one on a 300 m long rail at the end. In the end this was increased to ten and two when it turned out the budget allowed this. That gave twenty combinations between a fixed and a movable antenna and in twelve hours it could simulate a single telescope with a diameter of 1.5 km; although consisting of rings with gaps in between (see Figs. 11.8 through 11.10). The entire 1.5 km aperture could be filled in four times 12 h with the moving antenna in different positions along the rail-track. At 21 cm wavelength, this would give a beam of 22 arcsec east-west. In 1975 another two antennas were placed on a rail-track another 1.5 km to the east so that the baseline became twice as large and the beam became 11 arcsec.

Originally (in the days of the Benelux Cross) a location was planned close to the Dutch-Belgian border, but that was no longer necessary. The forestry

Fig. 11.10 Element dishes of the Westerbork Synthesis Radio Telescope. This photograph is taken halfway along the array looking east. Provided by ASTRON, Dwingeloo

Hooghalen was the ideal place. It is only 20 km from Dwingeloo. On the site was the former Westerbork concentration camp—inhabited by part of the Moluccan community—and there was enough space for an east-west line of 1.5 km and even for a possible expansion later to the east. For a long time Gijs van Herk made measurements of the stars to accurately determine the east-west orientation and the positions of the elements with respect to each other, a distance of exactly 144 m apart. The antennas had to be as identical as possible. For this part Ben Hooghoudt was project manager, assisted by young engineer Jacob Wilhelm Martin (Jaap) Baars (b. 1937). The Werkspoor company again did the structural design and construction. The electronics of the receivers were again developed, designed and built by Lex Muller with the cooperation of Jean Luc Casse (b. 1934), a Belgian electronic engineer who had already been hired during the BCAP phase. The online computer control was built and programmed under Ernst Raimond, who also became the first director of the Westerbork Radio Observatory. It was also vital that the data could be converted offline into a calibrated map of the distribution of the brightness of radio radiation on the sky. This work of developing the techniques and writing the computer programs was largely done by Wim Brouw. In 1968, the construction of the antennas was completed; the tests started in 1969 and lasted until 1970. For more background see [120] through [123].

Just before Oorts retirement, the telescope was put into operation. On June 24, 1970 the Queen landed on the grounds of the telescope to perform the opening act, see the beginning of Chap. 1. The investments through RZM's budget were of the order of twenty million guilders (40 million € nowadays). In all this, the support from ZWO and in particular Bannier was of great and decisive value.

11.4 Retirement

The first half of 1970, especially the month of April during which he had his seventieth birthday, was an extremely busy time for Oort. From April 13 to 18, there was a *Semaine d'Étude* at the Vatican about *Nuclei of Galaxies*, this time organized by Martin Ryle, Allan Sandage and Bill Morgan. Oort had been a member of the organizing committee as a member of the Pontifical Academy. In his contribution he reported extensively on the central parts of our Galaxy and the Andromeda Nebula, with a lot of attention for my work on the central regions of our Galaxy. The conference was characterized by a consensus that activity in the nuclei of galaxies was a common phenomenon. Radio astronomy played a major role in this. I already mentioned that the strong radio source Sagittarius A is in the core of our Galaxy. Virgo A appeared to be associated with the brightest galaxy in the Virgo cluster (M87), in which a 'jet' indicated that the core was active (see Fig. 6.4). But for example Cygnus A turned out to be a very distant, faint galaxy (it was first thought to be two colliding galaxies), which could only be seen with the Palomar 200 in. And Maarten Schmidt had discovered the quasars (initially called quasi-stellar objects), point-like objects in the optical at a very great distance, which emitted enormous amounts of radio radiation; although they looked like stars in the optical, they turned out to be the extremely bright nuclei of distant galaxies. Radio sources were associated with very energetic phenomena such as explosions in nuclei of galaxies. Margaret Burbidge, who summed up the conference, stated that there now was convincing evidence that nuclei could produce large amounts of energy and eject matter.

Fig. 11.11 (Continued)

This was followed by a symposium in the Netherlands to celebrate the Oort's seventieth birthday and the fact that the Westerbork Radio Telescope was ready to go into operation. This symposium on *Radio sources and the Universe* was held at the Kapteyn Observatory in Roden in the north of Drenthe. This observatory, part of the astronomy department at the University of Groningen, had been built to develop and construct instruments for the ESO optical telescopes in Chile (and later the British-Dutch ones at La Palma, Canary

Fig. 11.11 Group photograph of the attendees of a symposium *Radio sources and the Universe* at the Kapteyn Observatory near Groningen, on the occasion of Oort's 70th birthday and the inauguration of the Westerbork Synthesis Radio Telescope. From the Oort Archives. In the first place there were persons that had been very much involved (as astronomers or engineers) in realizing the Westerbork Telescope, both then and earlier on (with the numbers in the figure): Chris Christiansen (1), Ernst Raimond (4), Wim Brouw (6), Ben Hooghoudt (10), Jean Casse (16), Johannes Petrus Hamaker (b. 1937) (26), Bill Erickson (36), Kelvin J. Wellington (40), Lex Muller (41), Jaap Baars (42), Charles Seeger (45), and Jan Högbom (50). Old time (and more recent) collaborators and colleagues included Wallace Leslie William Sargent (1935–2012) (2), Lukas Plaut (3), Peter Scheuer (11), famous cosmologists John Archibald Wheeler (1911–2008) (13) and Martin John Rees (b. 1942) (18), Willie Fowler (22), Henk van Bueren (28), Otto Heckmann (29), Margaret and Geoffrey Burbidge (37, 49), Gart Westerhout (38), Harry van der Laan (51), Fred Hoyle (53), Lo Woltjer (55), Maarten Schmidt (56), Adriaan Blaauw (57), and Hugo van Woerden (58). Finally, students and young staff at the time, that would soon become early Westerbork users or related researchers, included Vincent Icke (8), Paul Ronald Wesselius (b. 1942) (14), Renzo Sancisi (b. 1940) (15), Ronald John (Ron) Allen (b. 1942–2020) (27), Jet (Katgert–) Merkelijn (31), Peter Katgert (32), Rudolf le Poole (43), Arnoldus Hendrikus (Arnold) Rots (b. 1946) (47), myself (Piet van der Kruit) (48), and Walter Joseph Jaffe (b. 1946) (52)

Islands). It was also a reunion of former colleagues and students of Oort (see Fig. 11.11). During this symposium Oort gave a well-attended lecture in the auditorium of the Academy building of the University of Groningen.

And then on April 28th there was the birthday itself with a big party with staff and students of the Sterrewacht.

In June this was followed by the official opening of Westerbork by Queen Juliana. After she had landed by helicopter and was shown around the control center, she pressed a button. The twelve antennas, all pointing to different directions, then moved to the same position in the sky. I have not been able to find anyone who remembers what that position was. Most of the audience, including me, experienced this from the large construction hall. Grandson Marc Oort only remembers that the overwhelming noise of the helicopter which made it impossible to hear speeches.

And on September 18 Oorts gave his formal farewell lecture (valedictory speech) in Leiden [124].[1] As was customary at the time, professors retired at the end of the academic year in which they turned seventy. On that occasion the Minister of Education and Science, Gerard Veringa (1924—1999), appeared, who pinned the very high Royal decoration of Commander in the Order of Orange-Nassau on his lapel.

In August Oort attended at the General Assembly of the IAU in Brighton, the fourteenth of both Oort and the IAU. The move that had to take place

---

For an English translation see Appendix B6 of [2].

Fig. 11.12 The house in Oegstgeest that the Oorts bought and where they moved to in 1970 when they had to leave the director's house at the Sterrewacht after Oorts retirement. Below: Oort in the garden of the house with the sundial presented to him by family, friends and colleagues. From the collection of Abraham H. Oort

was more drastic, because the Oorts had to leave the official residence at the Sterrewacht, where they had lived with an interruption in the war since 1935. They had bought their own house in Oegstgeest at a canal that was connected to the nearby lake district de Kaag. From financial contributions of many friends, family and colleagues from all over the world three gifts were offered. A painting, a sailing boat and a sundial in the shape of a radio telescope. Although Oort still went to the Sterrewacht on a daily basis, especially as the

Westerbork Radio Telescope began to produce fantastic results, the Oorts began a new phase in their lives. In addition, they spent more time in their country home in Haamstede on the island of Schouwen-Duiveland in the south-west of the Netherlands, which they had bought in 1960. Just like the country home in Katwijk, which was demolished by the Germans during the war to make way for the Atlantic Wall, they had named it Sandy-Hook (Fig. 11.12).

The question of his succession had been followed by Oort with great interest. Formally this was a task for the most senior of the other professors, Henk van de Hulst. Oort himself had put his hopes on his two students Lo Woltjer and particularly Maarten Schmidt. Meanwhile Woltjer worked at Columbia University in New York and had no real interest to leave. Schmidt was Oort's favorite; after all, with his expertise in quasars and radio sources he seemed the ideal man to lead the research in Westerbork. Schmidt hesitated, but eventually preferred to stay at Caltech, where he had easy and extensive access to the Palomar 200 in. He, and his wife and three daughters, were by now well settled in the American and Californian society. And a large ESO telescope was not really a realistic alternative to the Palomar 'Big Eye'.

The student protests of 1968 calling for more participation in the decision making process and the implementation of new structures at the universities had also changed the position of director. Eventually Henk van de Hulst took over the leading position within the Sterrewacht. Harm (Harry) van der Laan (b. 1936) was in charge of the Westerbork research. As a 16 year old he had with his parents emigrated to Canada (his father is said to have remarked, 'I have three sons and only one farm'), but had obtained his PhD in England with Martin Ryle in Cambridge with a thesis on supernova remnants as radio sources. In 1967 Oort had lured him to Leiden.

11.5   Westerbork and Nearby Galaxies

The primary reason for Oort to build the Westerbork telescope was to be able to perform counts of radio sources and undertake studies in cosmology and the structure and evolution of the Universe on a large scale. The first receivers at Westerbork were not for the 21 cm line (this would take a few more years) at frequency 1421 MHz (MegaHertz), but continuum radiation nearby in frequency (1415 MHz). The program of source counts would take a rather long time, because much had to be learned and determined about the instrument. To do that properly so that it could be used at the faintest levels, would take a lot of time. The first dissertation in this area, which was based on observations at Westerbork, was by Peter Katgert (b. 1944) and was defended only in 1977.

It took that long to fully understand all the observational problems to publish results with confidence. Eventually, Oort's grandson Marc Oort would also write a PhD thesis on this subject in 1987. The first spectacular results from Westerbork came from observations of nearby galaxies.

An obvious question was to see if nuclei of other spiral nebulae were also a radio source, and if they were whether the one in our own Galaxy (Sagittarius A) was extraordinarily strong or faint. That program ended up with me after I had finished my thesis research. I also observed some Seyfert galaxies, which have optically bright cores and in which gas moves at high velocities. The latter all showed radio sources, but the 'normal' systems generally did not. Our Galaxy was therefore not very exceptional given the luminosity of the sources that were detectable. Even before I was allowed to publish that result, Oort wrote a short article, so that he would be the first to present Westerbork-results in the literature. The Italian astronomer Paolo Maffei (1926–2009) had discovered two galaxies very close to the Milky Way on the sky, which could only be seen in the near infrared because of the absorption, but were very close to us. One of them was elliptical and Oort found that that one did not show a central radio source.

The first remarkable results came when we observed large, nearby galaxies. I was also responsible for this, but a visiting Australian astronomer, Donald Seaforth Mathewson (b. 1929), collaborated with me. Oort had to vacate his large office and, like other staff members, was given a much smaller office. His original office, which now had been assigned to Mathewson, still had his large collection of books along the walls, especially journals, and he came to look something up in it very regularly when Don and I were working on the Westerbork results. I strongly suspect that it was just an excuse to remain informed about our work and talk to us about it. Our first important result came with observations of the whirlpool nebula, M51. This system has an exceptionally well-defined spiral structure (and a companion galaxy which may have induced this by its gravity). The observations in Westerbork were spectacular (see Fig. 11.13), because we saw spiral structure. This was in fact a direct confirmation of the density wave theory. We saw that gas is slightly accelerated when it approaches the spiral wave pattern in the density of matter from the inside of an arm and slows down again when it has past it. Near the arm, the gas then runs into the gas that came in slightly earlier and has been slowed down a little. That mutual velocity is supersonic (the velocity of sound in the interstellar gas is of the order of a few km/sec) and a shock wave is created as a result. But when the gas is compressed, the magnetic field goes along (it is 'frozen in' to the gas because it represents much less energy) and the very fast and energetic relativistic electrons probably also. That gives

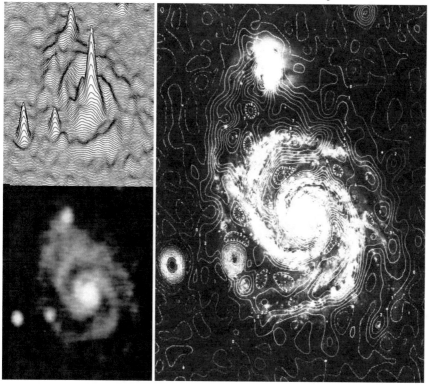

ig. 11.13 Westerbork observations of the continuous radio radiation of the whirlpool galaxy M51. Top left a representation of the intensity along horizontal lines (of decliation), bottom left a cathode ray tube image produced by Walter Jaffe. On the right ontours of equal radio brightness superimposed on an optical image. From a publica-ion by Mathewson, van der Kruit & Brouw in 1972. From the author's collection

nhanced synchrotron radiation. The dust is compressed along with the gas s well and that explains the dark dust lanes that are observed on the inside f the arms. Now the spiral structure that was seen in the radio radiation was ocated exactly on these dust lanes. This coincidence was a clear proof for the xistence of density waves. The compression in the gas is seen as the cause of tar formation, which then results in HII areas around newly formed O-stars little further downstream in the spiral arms.

This work was exactly what Oort was interested in and he regularly spoke bout it with great enthusiasm at international conferences. The remarkable hing was that this 'map' of M51 was created with a time investment in Vesterbork of no more than four times twelve hours. A wonderful crop of)ort's efforts. Oort turned out to be particularly fascinated by another obser-ation that Mathewson and I made, and that was of the NGC 4258 system. 'his is also a very bright galaxy—it appears in Messier's list as number 106; for

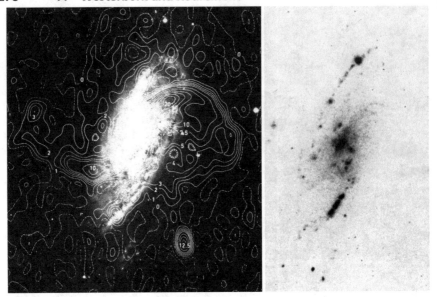

Fig. 11.14 NGC 4258. Left contours of equal radiocontinuum surface brightness super-imposed on an optical image. On the right the system in the light of a spectral line of ionized hydrogen (Hα). Figures from an article by van der Kruit, Oort and Mathewson in 1972. From the author's collection.

obscure reasons it is more familiar with the NGC (New General Catalogue) number. The observations showed radio emission from arms that do not correspond to the optical spiral structure (see Fig. 11.14). I soon found in the library that there was a French study in the light of ionized hydrogen (Hα), which showed 'anomalous' arms that coincided with the radio structures. When Oort saw this he became very excited and he modestly asked if he might participate in the analysis, which Don Mathewson and I gladly accepted without giving it any thought. Oort then collected from his bookcases all the bound journal volumes, which contained publications about this galaxy, and took that enormous pile with him to go through during a stay in Haamstede, for which he was to leave that same day. I assume he came by car that day and not as often by bike.

The possible explanation, published in an article with Oort as co-author, was that this was gas ejected from the nucleus much like seen in our own Galaxy in the inner regions, now not at an angle with but *in* the plane. The gas then makes its way as a plow through the existing gas and at the interface produces the shock waves by compression of the magnetic field and the cosmic rays in the disk and enhances the radio emission. The shock wave also ensures that the gas is not only compressed, but also ionized. Oorts comments given over the years, that spiral structure can occur induced by any disturbance of the disk, led to speculation that this was a way in which spiral structure might originate. Later a number of different explanations for the anomalous arms have been given, but none is able to explain all aspects of the phenomena in NGC 4258.

11.6 More Astronomical Research

The original hope was that radio source counts could be used to determine the structure of space. Einstein's Theory of General Relativity was the basis of the description of the Universe and the question was whether the Universe would continue to expand indefinitely, was critical and would stop only after an infinite time, or would rather start to contract again in the future. The curvature of space was the determining factor and this could be found from the run of the number of sources as a function of their brightness in the sky. If sources could be identified with a small galaxy, then the redshift and thus the distance could be measured through the radial velocity (the redshift of the spectral lines). Jet Merkelijn had spent some years in Australia after her doctoral (masters) exam and had studied optical identifications of southern radio sources. Back in Leiden she turned that into a dissertation, on which she obtained her PhD with Oort as supervisor in 1970. The important discovery was that the 'luminosity function', i.e. the distribution over intrinsic luminosity, changed with the distance from us and thus with cosmic time. This made the interpretation of radio source counts more complicated; at the end of the day the censuses was that radio source counts give more information about the evolution of the population of radio galaxies and quasars with the age of the Universe, than about the structure of the Universe itself. Oort continued to follow these developments, but did not do any further research in this area; the work in Westerbork was led by Harry van der Laan.

After his retirement Oort still came to the Sterrewacht on a daily basis, on his bicycle from Oegstgeest. At first he set to finish work that had been waiting for completion for some time. The first was a project that Gerrit Pels had started years before concerning the membership of the closest star cluster Hyades. This cluster covers a large area in the sky; the brightest stars of the constellation Taurus (except the brightest one, Aldebaran) are part of it. The cluster is at 48 pc or so from us. The Pleiades, about which Oorts PhD student Leendert Binnendijk had written his thesis, are not far from the Hyades in the sky, but they have a distance of 135 pc. In Greek mythology, the Hyades and Pleiades are daughters of Atlas. Because the Hyades are so close, they are a good springboard for distance measurements. Pels had determined many proper motions in the field to see which stars belonged or did not belong to the Hyades, but became ill and died in 1966. Together with Heleen Kluyver, who was married to Pels, Oort finished the work. Also in the Hyades the heavier stars are more concentrated towards the center.

Lukas Plaut had been working on the Palomar-Groningen Variable-Star Survey since 1953, which Baade had proposed at the Vosbergen conference. That work was completed and published in the early 1970s and Oort and Plaut together interpreted the results in terms of the distance to the Galactic Center by determining where along a line of sight through Baade's Window the density of variable RR Lyrae stars was highest. The distance was thus determined to be 8.7 ± 0.6 kpc.

Another piece of unfinished work was the never ending story of NGC 3115, which Oort already discussed as far back as in 1939. The dynamics never became clear; in spite of better observations of the rotation curve, problems remained with unrealistically high values of the mass-to-luminosity ratio. The Oort Archives contain an number of draft manuscripts for publication through-out the 1950s up to the 1980s. In the mid-1970s, Oort had asked Maarten Schmidt to try again with then existing instrumentation to take a new spectrum at the Palomar 200 in., which he did and then had it reduced by a student at Cal-Tech. This resulted in a much lower rotation velocity at a larger distances from the center than all previous measurements. This was finally the reliable result that Oort had been waiting for for decades. But he never got around completing an article on this subject.

For many years Oort had good contacts with the couple Margaret and Geoffrey Burbidge, who in the 1960s regularly visited Leiden for a few weeks in the summer. Geoffrey was editor in chief of the series *Annual Review of Astronomy and Astrophysics*. This is an annual publication in book form, in which scientists are invited to address a particular subject and discuss recent progress and the state of affairs. It is considered prestigious to be invited to do so. In 1974, Burbidge had asked Oort to write an article about the central parts of the Galaxy. Oort hesitated, because he still had a lot in the pipeline (among others NGC 3115, as he wrote to Burbidge). But Burbidge did not give up. Oort took the bait and promised to make time for it in the summer of 1975. But he did not finish the article in time and promised to send a manuscript a year later (deadline October 1976). He did not make that either; in November he sent a first installment and later some more until finally the manuscript was complete in February 1977. Oort often took things too lightly and underestimated the time it would take him. The final article was 57 pages long (while, as is the standard, 20 had been allocated). It is a miracle that

Burbidge managed to get it processed and fitting in on time for the volume that appeared that year, but he did.

It was an excellent piece of work. The expansions at a few kpc (where my dissertation was about) were no longer attributed to ejections of matter from the nucleus, but closer to the core there were all kinds of phenomena such as gas clouds moving away from the nucleus and radio and other energetic radiation, which indicated a large concentration of matter of a few million times the mass of the Sun within a pc or less. Oort suggested the presence of a black hole of that mass. Since then, these clues have only grown stronger.

In 1974 the staff of the Sterrewacht Leiden moved to a new complex of institutes and laboratories on a campus on the north side of Leiden (Figs. 11.15 and 11.16). The astronomy department was housed in a tall building on the Wassenaarseweg, the Huygens Laboratory, and Oort of course moved along. Naturally he did not have enough space there to put all his books and journals in his office and his house in Oegstgeest was not an option either. He decided to donate his collection to the library of ESO, the European Southern Observatory of which Lo Woltjer was now Director General. For the time being Oort only donated part of his collection, but he promised that the rest would follow when he no longer was actively involved in research. Woltjer accepted it in January 1976. ESO had at that time rented some buildings from

Fig. 11.15 Oort in 1984 with the Huygens Laboratory in the background, to which the Sterrewacht had moved in 1974. The building has eleven floors and the Sterrewacht was mainly located on the sixth and seventh floors (American notation). From the collection of Abraham H. Oort

Fig. 11.16 Oort in 1980 in his office at the Sterrewacht after it had moved to the Huygens Laboratory on the Leiden campus on the north side of the city. From the Oort Archives

CERN in Geneva (while the Director's office of ESO was based in Hamburg), but the intention was that when ESO went to a permanent housing, a special place in the library would be reserved for the Oort collection. That new building was opened in Garching bei München in 1981. The remaining part of the Oort collection followed in 1982, although Oort was still conducting active research. Incidentally, in 1998 the Sterrewacht, as the name of the department of astronomy remained to be, moved into a new building just next to the Huygens Laboratory, which was christened the 'Oort building'. It has a special architecture in the sense that it slopes some ten degrees out of the vertical.

11.7   Pocket Diaries and End-of-Year Highlights

The Oorts celebrated their 40th wedding anniversary in 1967. On that occasion Mrs. Oort gave her husband a plaque, which is shown in Fig. 11.17. It is a quote from the poet Pieter Corneliszoon Hooft (1581–1647): 'I devote myself to your service, my love, my light, my life.'

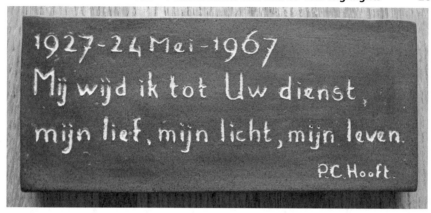

1927-24 Mei-1967
Mij wijd ik tot Uw dienst,
mijn lief, mijn licht, mijn leven.
P.C. Hooft.

Fig. 11.17 Plaque that Mrs. Oort presented to Oort on the occasion of their 40th wedding anniversary in 1967. For explanation see the text. From the collection of Abraham H. Oort

From 1964 onward, Oort was in the habit of collecting special items to list in his pocket diaries at the end of the year. There was also a list of books he had read and ones he and Mrs. Oort had read to each other. Somewhat later he started listing the meetings he had attended and the trips he had made, and he commented on his students and on his research. And, increasingly, the births of grandchildren and their progress in school, as well as the deaths of family members and friends.

To give an idea: according to the pocket diary for 1965 they went to their second home in Haamstede 11 times and spent 65 d there (Mrs. Oort even spent 8 more). He also wrote 'to have skiffed quite regularly'; he continued rowing until a very old age. Yet at the beginning of this diary he had written an overview of 1964, which ended with 'At the end of the year I am tired and feel old'. About 1965: 'I turned 65 and joined the A.O.W.'. AOW is the state pension every inhabitant of the Netherlands receives after the age of retirement (then 65). 'My own work went very poorly. Still the article on NGC 3115 is not finished, nor that about the high-velocity clouds. In the last 4 months of the year, however, I have been working on the latter again with enthusiasm'. There has been a 'Big Sterrewacht party on Dec. 17', and 'Lustrum of the University, [390th anniversary] with big closing party on 't Rapenburg and performance of Tijl Uilenspiegel on 't Pieterskerkhof', 'Bertil Lindblad died on 25 June'.

In Haamstede for many years there was no television and telephone to keep daily life nice and quiet. But in 1969, when the first Moon landing took place, Oort hired a TV set to watch the first steps on the Moon. The quiet lasted until Mrs. Oort won a television set in a lottery and there was no better destination

Fig. 11.18 The Oorts with Oort's brother John and sister-in-law Pivy. This picture dates from 1984. From the collection of Abraham H. Oort.

for it. It disappeared after a few years in a burglary, but then the tone was set and a new one was purchased. Oort's brother John and sister-in-law Pivy (see Fig. 11.18) also often visited Haamstede. In 1968 John retired as a professor in Wageningen and he and his wife also bought a house in Haamstede.

In January 1974 Mrs. Oorts mother Hélénè Graadt van Roggen (in the family grandma Lène) died. She was 93. Her son had died in 1933 and her husband in 1945. At her funeral Oort said (written together with Mrs. Oort and made available by Bram Oort):

'A new emptiness came into Mother's life when father died towards the end of the war and many relationships and friends disappeared from her life with him. Some time later, however, a new family center began to form in Utrecht and De Bilt. First Coen got a job at the Jaarbeurs [... and later he became a] professor in Utrecht and [...] eventually Bram and Bineke settled in De Bilt. Meanwhile, with a decisiveness with which she completely beat us all, even Coen, she had decided in a single day to put her house [...] up for sale and buy an apartment in Den Akker. How happy she has been with her 'own' apartment. [...]

How nice it would be if grandma could meet her husband and her son Coen again and they could catch up together. It was not in Grandmother's nature to delve into such speculation. Only once she said something like that, during Grandpa's funeral service on April 21, 1945, in which none of us could be present because of the war. She said, among other things: 'Dying is not like a cup falling to pieces and dying is not like a cord snapping and the beads getting lost: but dying is: from the small stream flowing into infinite waters; 'the merging of the detail with the general'.'

In the pocket diary for 1979 (Wednesday 3 January):

'On the canal people skate in spite of the snow. I take out my skates to check out the ice. It is difficult because of the snow, but I can skate under the bridge and under the railway bridge to the Leede, but on the Leede skating is almost impossible. On the Dwarssloot [also named the 'Groote Sloot'] the snow has been blown off the ice and skating is nice, although every 15 m you have to step over a ridge of snow. On the Zijl it is crowded with skaters and the ice is reasonable. Just as I was about to return I fell on the back of my head and only woke up when I was being helped to the bank of the river supported by a female student, Sytske Bant. Standing on the bank I had a strange 'dream' experience, in which I first thought it was unreal that there was ice and that I had put on my skates at home. After 10 to 15 minutes the girl's boyfriend, Mr. Groothuis, came to take me home in his car, where I arrived at 4:30 pm. An enormous mass of blood under my ice cap and with primitive bandaging I go to bed. Henk Bannier is coming to dinner. At 7:00 pm physician Held arrives and puts a large plaster bandage on my head. He orders me to keep to my bed for 4 or 5 d and have few visitors. I do talk a bit to Henk anyhow, but retire with [a fever of] 38.5.'

Skating to the small river 'Zijl' is about 5 km, but Oort did not record how far he skated on it, but in total he probably had skated no more than 10 km or so before deciding to return. The next day Oort took a rest, but still read a few articles and wrote two letters to the authors. On Saturday (January 6) physician Held came to check up and Harry van der Laan to visit. Harry came on skates (he lived some 15 km skating away) and told about a car accident of his son Vincent and the funeral of his father in Canada. The following Sunday was Mrs. Oort's birthday and Coen and Marijke with their families visited and they sat near the fire place, except for Coen who left after lunch as he had to travel to Mexico (by Concorde, the supersonic airliner). Oort noted that at the end of the day he was rather tired. On Monday he still feels a bit light in the head, but seems OK again on Tuesday.

Oort never skated by himself again; Mrs. Oort had ruled that they would only skate together, which they did occasionally. However, he kept up rowing until he was about ninety, but stopped using his sailing boat at about eighty because it involved so much effort to prepare.

JAN HENDRIK OORT

ASTRONOMIAE LECTOR, POST PROFESSOR 1930 - 70 ·
NATUS FRANEQUERAE 1900 · OBIIT 5 · XI · 1992
IN PAGO WASSENAAR

Fig. 12.0 Painting of Oort that is displayed in the Senate Room in the Academy Building of Leiden University. It stems from around 1980 and has been painted by Gerard de Wit (1931–2010). From the collection of the Academy Building, University of Leiden, see [125]. Reproduced with the permission of the de Wit family

12

The Last Horizon

It is terrible to become eighty,
but not to become eighty is even more terrible.
Harry Mulisch.

At the end, the horizon is reached;
it turns out to be incomprehensible.
Jan Hendrik Oort

Oort's eightieth birthday (Fig. 12.1) was marked with the publication of an *Liber Amicorum*. This 'book by friends' was entitled *Oort and the Universe: A sketch of Oort's research and person* and was edited by Hugo van Woerden, Wim Brouw and Henk van de Hulst [1]. Several friends, colleagues and former students contributed to it. I quote from the chapter by Henk van de Hulst on style of research':

> ... the best antidote to perfectionism: to remain open to and get inspired by an even more exciting problem. Oort manages to do so quite well and this point alone, a very genuine interest in the marvelous Universe, may very well be the key to his style of research.

Van de Hulst further characterizes Oort with two features, his 'careful treatment of observation material is the root of astronomy' and 'basically he remains an avid observer'. Indeed, Oort was not a real theorist. The foundations of Galactic Dynamics were laid by James Jeans and Arthur Eddington, not by

---

The quote from Oort is from the cover letters he sent to friends and colleagues presenting them a reprint of the *Zenit* issue with the Dutch translation of his Kyoto Prize Lecture.

© The Editor(s) (if applicable) and The Author(s), under exclusive license
© Springer Nature Switzerland AG 2021
P. C. van der Kruit, *Master of Galactic Astronomy: A Biography of Jan Hendrik Oort*,
Springer Biographies, https://doi.org/10.1007/978-3-030-55548-1_12

Fig. 12.1 Oort in 1980, probably at the celebration of his eightieth birthday. From the Oort Archives

Oort. He applied their insights to the observations of stars and thus discovered the rotation of the Milky Way Galaxy and formulated a description of the properties of the distribution and motions of stars in the solar neighborhood. As Oort himself said in an autobiographical article in 1981 in the *Annual Review of Astronomy and Astrophysics* about his great example Kapteyn [126]: 'Two things were always prominent: first the direct and continuous relation to observations, and secondly to always aspire to, as he said, look through things and not be distracted from this clear starting point by vague considerations.' Van de Hulst also made the following point:

> Oort once said: 'you can do only one thing well at a time'. I think he follows this maxim himself. But he has managed to do many things well at different times. In this respect his style of work must be classed as somewhat nomadic, although he returns very frequently to his preferred pasture, the structure of our Galaxy and it mysterious center.

12.1 Superclusters

The structure of the Universe on a large scale was Oort's last major interest. The counting of radio sources and its application to cosmology, the determination of the structure of space, led to the construction of the Westerbork Radio Telescope. Already in 1964 he was asked to write an review article on this subject for the *Annual Review of Astronomy and Astrophysics*. He did not have the time, so in 1967 Martin Ryle was asked to write such a review, but in 1967 Oort was asked again to cover this subject as soon as the data from Westerbork came in. This took too long and in 1971 Oort returned the request.

However, Geoffrey Burbidge, the editor-in-chief, insisted on Oort's contributions to the *Annual Review*. This resulted in his comprehensive and much quoted chapter in 1977 on *The Galactic Center*, which has been discussed above. But Burbidge did not stop there. In 1979 he invited Oort to write an autobiographical chapter; every year the volume opened with such a chapter. Oort hesitated but did write it [40]. 'I am not good in history, have no vivid memory of the past, and am much more interested in the future. I wish the Editorial Board had thought of a, to me, more intriguing subject (such as, for instance, the prospect of measuring the curvature of the Universe)'. Burbidge

Fig. 12.2 Oort with Pope John Paul II, Karol Józef Wojtyla (1920–2005), during the *Semaine d'Étude* at the Vatican in 1981. From the Oort Archives

replied that if he first did the autobiographical piece, he could 'then go on and write on the topic that you're are most interested in. I shall be delighted to publish whatever you send us.' Oort took on the task and wrote the requested autobiographical chapter for the volume of 1981, after which he promised to write about superclusters.

Superclusters are the largest known structures in the Universe, sometimes up to ten times larger in diameter than the 'normal' clusters of galaxies. They have an irregular shape, contain several of those regular clusters and have crossing times (the time it takes for a galaxy to go from one side to the other), which is longer than the age of the Universe. This explains their irregular structure as being too young to relax. A typical size is 350 Megaparsec (Mpc) and they have a typical mass of $10^{15} - 10^{16} M_\odot$ (something like ten thousand times the mass of our Galaxy). In a way Oort returned to his fascination with the distribution of galaxies, with which he had opened his inaugural speech. Oort started his chapter on superclusters in the *Annual Review* with the same figure about the distribution of bright galaxies over the whole sky as in that speech in 1936.

Again it took a long time before Oort had written the article. The subject fitted well in a new *Semaine d'Étude*, which was held at the Vatican around October 1, 1981, this time on *Cosmology and fundamental physics* (See Fig 12.1). There Oort presented a lecture on 'The largest structures in the Universe'. He did not meet the deadline for the *Annual Review* (October 1981) and in December he sent a first part followed by the rest in January 1982. But Burbidge this time had no room left in the 1982 volume, so Oort rewrote the whole thing. Again he was too late for the deadline, but in December 1982 he managed to get a manuscript to Burbidge. It appeared in 1983 and again it was much longer than the usual twenty pages, namely 56 pages. Apart from a beautiful overview and discussion of all observations that appeared in the literature, Oort concluded that superclusters were formed as flat structures 'pancakes', as a contraction in the first gaseous stage of the Universe even before the formation of the galaxies.

It became more and more difficult for Oort to go to the Sterrewacht. In 1987, when he could still come regularly, he said in a newspaper interview [127]:

> It is not easy to go through all the astronomical journals every month. But apparently at a very advanced age one can still remain curious, have the urge to find out many things. I still concern myself with the largest structures in the Universe. It is amusing to conclude that in the 1920s my borders were determined in space, by those of the Milky Way Galaxy that were at 20.000 lightyears. Now, my horizon is determined in time, 15 billion years, the age of

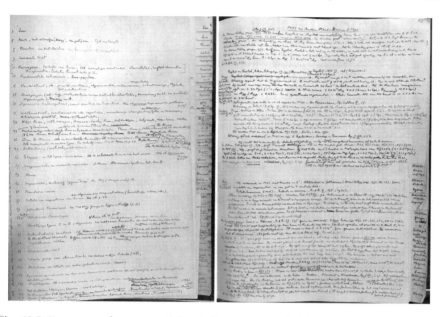

Fig. 12.3 Two pages from one of Oort's literature notebooks. On the left a table of contents; note the cut-out tabs on the right. On the right a sample page (not intended to be readable). From the Oort Archives

the Universe. It is a pity that before too long I will have to die and then can no longer see the new developments.

The growing volume of scientific publications Oort is alluding to in this quotation is reflected in every astronomer's increasing difficulty of keeping up with the literature. This raises the question of how Oort coped with this. Oort's Archives contain seven hardcover, large-size (varying, but of order 20 by 35 cm) notebooks, of which two sample pages are shown in Fig. 12.3. The pages have been cut on the right-hand side to provide tabs for browsing. Each tab corresponds to a general area ranging from Solar System to cosmology. There are notes for each article read in Oort's neat but tiny handwriting, usually restricted to up to four or five lines per paper. Although he read articles in English, the notes are in Dutch.

Oort was evidently very conscientious in keeping up with the literature. The earliest entries are from 1945, the latest from 1983, after which he must have given up following the astronomical literature systematically. It should also be mentioned that for a long time he had personal subscriptions to major journals, which were bound per volume in hardcovers. These occupied many meters of shelf space in his office at the Sterrewacht. These went to the ESO library.

12.2 Return of Halley

When in 1986 Halley's comet returned to the inner parts of the Solar System, Oort's theory of the origin of comets and the Oort Cloud received a lot of public attention. Originally Oort had thought, that he had also seen Halley's comet during the previous passage [128]:

> At the last passage of Halley 1910 comet, I did see a comet. My father, who was a physician in Oegstgeest, took me outside one evening and there was THE comet. I remember a beautiful tail. Later I realized that my father had no idea about astronomical matters. It was dark and cold early that night. However, Halley's comet appeared only in May in its full splendor in the sky and could be admired around midnight. Something was not right and over the years I became convinced that in 1910 I had not seen Halley's comet, but the Johannesburg comet, which was passing along the starry skies in the first months of that year.

That was the 'Great January Comet' of 1910, also called the 'Daylight Comet'. As the name says, it could be seen before the Sun set.

The public fascination for the comet also led to the intitiative by the Volkssterrenwacht Simon Stevin (a general public observatory) near Breda to organized seven flights with an Airbus-210 for 240 passengers, during which the comet would be visible at cruising altitude of 10 km. Oort was invited to join one of these flights (see Fig. 12.4).

The media also paid a lot of attention to the close fly-by of Halley's comet by the Giotto spacecraft launched by the European Space Agency ESA. It was named after the medieval painter Giotto di Bondone (1270–1337), who was probably the first to depict the Star of Bethlehem as a comet on one of his frescoes. Giotto was part of a set of five spacecraft, the 'Halley Armada' launched by various space agencies. Giotto would be the first to pass very close to the core and the results would be used to better determine its position for the later ones. The Giotto fly-by took place on March 14, 1986, in the Netherlands just after midnight. Dutch television broadcast a live report and Oort was first interviewed from the studio before the program was transferred to the ESA Darmstadt flight center. The interview was hilarious in my memory because to every question journalist Petrus Lambertus Lucas (Piet) Smolders (b. 1940) asked about the Oort Cloud, Oort gave some obligatory response but changed the subject all the time to the cosmology and large scale structure of the Universe, a subject which apparently occupied him more at the time.

Oort was stimulated by Halley's return to take another look at the statistics of the parameters of the orbits of new comets. The organizers of the Halley Lecture thought it appropriate to invite Oort for a second time (he had already

Fig. 12.4 'Destination comet Halley'. Oort is escorted by two stewardesses to board a Martinair charter flight to view Halley's comet from an altitude of 10 km. This took place on 9 January 1986. From a newspaper clipping in the Oort Archives, [128]

given a Halley Lecture in 1951). Oort agreed and revisited his theory. The update of the original orbit parameters had not yielded any new insights.

12.3   Balzan and Kyoto Prizes

Oort received the Balzan Prize in 1984. This prize is awarded annually for exceptional achievements in four areas, including natural science. The prize is named after Eugenio Balzan (1874–1953), an Italian newspaper magnate. His daughter, Angela Lina Balzan (1892–1956), who inherited his estate, left instructions after her death to establish a foundation to award prizes. The first went to the Nobel Foundation in 1961, and a few individuals received the prize in 1962. But then disagreement arose in the foundation about, among other things, the appointment of new board members. The chairman and executor of Angela Balzan's will, Father Enrico Zucca (??–1979), a Franciscan, acted too arbitrarily according to the others. The governments of Switzerland and Italy (the headquarters were in Zürich and the office of the Board in Milano), blocked the funds and dismissed the board, also—according to some—because more money was given away than came in as interest on the assets. It went so far that U Thant (1909–1974), on behalf of the United Nations, came to collect the prize in 1964, but received what turned out to be an uncovered cheque.

In spite of all this, Father Zucca sent letters to prize winners in 1965, including Oort and economist Jan Tinbergen. Of course this was never formally awarded. There was a many years long legal battle, about which, by the way, the website of the foundation is anxiously silent. It was not until 1978 that the prize was awarded again (to Mother Teresa of Calcutta). Eventually Oort was again nominated and selected for the prize in 1984. Tinbergen never got it (he lived until 1994 so it would have been possible), but he had already received the Nobel Prize for Economics in 1969.

The price was 250,000 Swiss Francs (in current purchasing power about 215,000€). Oort accepted the prize in Rome in November 1984. In his speech he said among other things:

> The pleasure of receiving this is particularly enhanced by the fact that it comes from Italy, the country where Galileo made his great contribution to our insight in the vastness of the world in which we live, [...]
> On the occasion of a high award one's thoughts naturally turn to the sources of one's scientific interest. The main inspiration in my life has undoubtedly come from Professor Kapteyn in Groningen, who was one of the earliest explorers of the Milky Way System. [...]
> In recent years my work has carried me to the largest existing structures and their evolution in the course of time. I wonder sometimes whether this work would have the approval of my former professor, strongly interwoven as it is with its speculative aspects. But on this point I derive comfort from a verse by Browning [Robert Browning (1812–1889)]: 'But a man's reach should exceed his grasp, or what is heaven for?'.

In 1987 Oort was awarded the Kyoto Prize. The news came, Harry van der Laan told me with a phone call from the Japanese embassy to the Sterrewacht while the Oorts were in Haamstede. Harry would tell them when they were back at their home.

> So they were in Haamstede and I received a phone call from the Japanese Embassy [...] and they told me Jan had won the Kyoto Prize, I think half a million Guilders or so. I knew they would return in two days, so I thought 'It is no use bothering them now, they will be packing for the return trip anyway'. So I called them in Oegstgeest [...] and Mieke answered the phone. I told her I had something to tell them but preferred not to do that over the phone. [...] When I arrived they had tea prepared, and for me it was only a quarter of an hour by bike. 'Well Harry', Mieke said, 'what is it you have to tell'. I said: 'I had a call from the Japanese Embassy'. She said: 'Jan, what do you have to do with Japan?' 'Well', Jan said, 'I have a few friends there, but nothing special.' And I said: 'According to the ambassador, or his spokesman, you have been awarded a major prize. It is called the Kyoto Prize and it means a large sum of money,

about half a million guilders.' Jan just looked, and Mieke said: 'Oh, then we can finally have this leaking roof repaired.' [...] I said: 'There is a complication in the conditions and that is that you have to pick up the prize personally.' Now Mieke sat in a wheelchair, because of problems with her knees. 'Well', Jan said, 'That should be possible. [and to her] And you have to come along.' Coen [their older son] worked for the ABN Bank and also was chairman of the Supervisory Board of KLM. So he arranged that KLM would sponsor this. So, Coen, his wife, and Jan and Mieke, went to Tokyo first class.

First there was a workshop, where Oort gave a lecture on the creation of structure in the Universe. He also held the Kyoto Lecture, *Horizons*. The text, which was originally in English, has only been published in Dutch in the amateur magazine *Zenit* [129][1]. A transcript from the actual notes Oort used in delivering the lecture in Kyoto, which of course was in English, has been included in the Appendices to my scientific biography of Oort [2]. It starts as follows (note the similarity to the first lines of Oort's first publication in 1922 on the occasion of the retirement of Kapteyn, quoted in Sect. 2.3):

As children we watched ships on the beach disappear as they sailed further and further away. We first saw the ship itself disappear, while the masts remained visible. As they sailed further, the tops of the masts also dived under the horizon. As children we fantasized about that horizon. What was behind it? Adults had the same curiosity. They sailed towards the horizon and saw that there were new horizons behind it and that this was repeated almost eternally. Until finally they realized that they had sailed around the Earth and discovered the whole world. But there remained space above us: the realm of Sun, Moon and stars. From the earliest times from which writings have come to us, it was thought that the celestial bodies formed the outer boundary of the Universe.

Oort distributed reprints of the article among friends and colleagues and the cover letter said:

Enclosed I send you an issue of Zenit which contains the Dutch translation of the lecture I gave in Kyoto. At the end the horizon is reached; it turns out to be incomprehensible.

With the help of Harry van der Laan and others, Oort and Mrs. Oort decided, after consultation with the children, to set up a foundation with the money from these prizes (the Vetlesen Prize had been used for the house in Oegstgeest) to finance a special visitors' chair, the Oort professorship. The Oort professor comes to Leiden for a few weeks, there will be a one-week workshop

---

For the English text see Appendix B7 of [2].

and an Oort lecture for a wider audience. The appointment is annual and began in 1990 with Lo Woltjer, followed in 1991 by Maarten Schmidt.

12.4 Personal Matters

Family gatherings were important for the Oorts and their family. A Dutch tradition is 'Sinterklaas' on December 5, referring to the Saint Nicholas (270–343), a Greek bishop of Myra (which actually is in present-day Turkey). In the folklore he comes from Spain and gives presents to children. Grown-ups present each other gifts wrapped in special, personalized ways (called 'surprises' from the French) accompanied by a poem with personal messages. To decrease the workload often lots are drawn so that one only has to prepare a gift for a single person. From notes provided by Marc Oort, grandson and astronomy PhD:

> Sinterklaas indeed was always a highlight. Lots of poems and surprises (no draws, just everyone made something for everyone else) and it regularly required working until after midnight. And no one was spared; the lesser qualities came on the table without restraint. Grandpa and Grandma felt it was a great party. They had two sessions; first with us and usually the day before or after with aunt Marijke. Any excuse to get the whole family together was allowed. They wanted to have everyone around them as much as possible.

In Fig. 12.5 we see the Oorts with the sculpture of John Rädecker's 'Hinde' (see Sect. 7.7) in the middle of the Kröller–Müller park in the Hoge Veluwe. This picture was taken by Bram Oort when they visited it and had a picnic there. Bram also wrote about his parents:

> When it comes to emotions, Mother was most in touch with her emotions. Father often did not seem to know what the 'fitting' emotions were in tragic events (he often looked at Mother to see what was going on). Maybe that was also because of his severe hearing disorder. Father was interested in everything that was new to him; he was 'naturally curious'. I remember an occasion when Father and Mother visited us in Rocky Hill; they had an apartment at the Institute for Advanced Study in Princeton. Bineke and I and our children were working on EST [Erhard Seminars Training]. Our eldest son, Pieter Jan, led an EST group on communication and invited the four of us to participate. Father was sitting in the front row (eventually Mother stayed home). I do not know how much he understood, but his reaction was how surprised he was that volunteers from among the participants could express their (deep) emotions so easily in public. After that he listened at home with attention to audio tapes of Werner Erhard, the EST guru at the time.

Fig. 12.5 The Oort near the sculpture of John Rädecker's 'Hinde' in the middle of the Kröller–Müller park. The Oorts owned a preliminary study of this in bronze. From the collection of Abraham H. Oort

In 1977 the Oorts celebrated their fiftieth wedding aniversary. I quote from the speech that Oort gave during the celebration with his wife and children, etc. (from the Oort Archives):

> My life may have been too full of my profession and always too many daily duties that I never finished in time, so I did not always enjoy the family consciously enough. That home was mother's creation: without any fuss, without asking credit for it [...]. But it was she who, almost unconsciously, created the family and our circle of friends at the Sterrewacht and the University, as well as our many guests, and built up the captivating experience of our lives. Intuitively feeling what is beautiful and what is good and always thinking of others more than of herself. [...] In the beginning (but also later) we talked about the ideals we wanted to pursue in life. In mother's case, it was simple: the desire to make something pure of it. With me, it was more complicated and vague and unattainable.

In 1987 at their sixtieth wedding anniversary Coen Oort said (made available by Bram Oort):

> In the foreground and the background of my early memories, your constant warmth, the great example (although I have often protested against it!) and still

keeping the family together. [...] You were very 'modern' parents, very tolerant, which we did not appreciate and understand until much later when we were raising children of our own.

It must have been a milestone and a source of pride that grandson Marc Oort (see Fig. 12.6) defended an astronomical dissertation in 1987. In his speech (in the Oort Archives) Oort said:

It was a good time those years together at the Sterrewacht. Years in which we both worked on cosmological subjects and were interested in the same things, even though our actual daily work was different. It was not until I read your thesis that I realized how much computer work and discussion went into it and how hard you must have worked. [...] Compared to you, I have an easy job: no observations to do, no computer to check and new commands to enter. Just reading literature and trying to summarize and understand. I liked the idea that you would continue the work and that you would be able to experience the wonderful things in astronomy in the next 5 to 10 years and that all this, so to speak, would 'stay in the family'. Unfortunately the job market in astronomy is so unfavorable at the moment so that there is no job to be found for you in it, so we will not be attending the colloquia sitting together in the front row in Room 2 much longer.

Fig. 12.6 Oort with his grandson Marc in 1984. Marc obtained a PhD in astronomy in 1987. From the collection of Abraham H. Oort

Marc Oort writes:

Grandpa came by bike to the Huygens lab almost till the end. He fell off his bike one day when he was 89 and was taken to the emergency room. The doctor asked (to check if he did have a concussion) what 7×7 was and immediately got the right answer. The follow-up question was 'and 7×17', and he also got an immediate answer. After a lot of thinking and calculating, the doctor came to the conclusion, to his surprise, that that answer was also correct. No concussion then (but several abrasions). Just like skating, he then (almost) stopped biking.

When I came to the Sterrewacht after my candidates (bachelors) it became a tradition that I would have dinner with him every week after the Thursday afternoon colloquium, so we biked to Oegstgeest together. Rule (Grandmother's) was that talking astronomy was not allowed then. He had a lot of difficulty with that and it rarely worked out. We kept this up for at least three years (then I got a girlfriend and shifted my priorities). Because our family is quite informal (always the informal 'you' to Grandma and Grandpa and always saying things the way they are) I have always benefited much from our time together at the Sterrewacht. He always told me when I was wrong or being sloppy (at my thesis defense he asked a question that came down to the fact that he found a certain conclusion absolute nonsense) and vice versa. I always joked with him. He was very sorry that the postdocs and PhD students always treated him with such respect. He would have preferred to sit at the lunch table with them and talk on an equal footing. But it rarely worked and he missed that. Especially after his hearing deteriorated. We shared an office for two months (after my PhD until I got a job at Fokker Space). His was the tidiest desk in the Universe, mine the most chaotic.

My thesis defense was indeed a unique experience. My uncle Tom (who as a professor of law had also talked himself into the committee) recalled that there were 5 generations of Oorts present. From Henricus Oort (in a painting on the wall) to our (then unborn) eldest son. Afterwards I heard that he had been thinking for days about how he would address me. Eventually it became the neutral 'mister candidate'.

The private correspondence in the Oort Archives shows a great concern with humanitarian affairs. In the 1980s Oort signed several petitions requesting the release of political prisoners in Chile, Turkey and the USSR. He wrote personal letters to Chilean leaders including President Pinochet (in 1987) and the Prime Minister and President of Turkey (in 1988). He was concerned about the treatment of Russian dissidents such as physicists Andrei Dmitrievich Sakharov (1921–1989) and Yuri Fyodorovich Orlov (b. 1924). Like others, however, he did not give up his membership of the Russian Academy, because he thought he might have more influence as a member than as an ordinary world citizen. Both physicists have been released.

Increasingly Oort complained in letters and conversations about the deterioration of his hearing. In October 1986 he wanted to see me before a meeting of the Westerbork Program Committee, which discussed observing proposals for the WSRT and granted observing time. He had been a member since the beginning and as an honorary member he never needed to be reappointed. He appealed to me as chairman of that committee with the announcement that he wished to resign. The reason was that he was absolutely unable to follow the discussions. I asked him if he still wanted to write reviews of proposals for the committee, but he preferred to quit altogether, even though he really regretted that. He also did not return for one last time, as I suggested, for a formal goodbye. He had seldom missed a meeting.

12.5   The End

In 1990, on the occasion of Oorts 90th birthday, a large reception was organized, attended by old and current friends, colleagues and family. The venue was Castle Duivenvoorde in Voorschoten near Den Haag. The castle dates from 1226 and is therefore one of the oldest castles in Holland (the western part of the Netherlands). During this reception a picture was taken of Oort with three Dutch Directors General of ESO, see Fig. 12.7; two former, Adriaan Blaauw, Lo Woltjer and the then incumbent Harry van der Laan.

In the summer of 1991 it became impossible to live by themselves in Oegstgeest. As grandson Marc Oort told me, there had been several nasty accidents, like one time when one of the two had fallen and could not get up on his/her own anymore. The other wanted to help, but also fell and could not get up either. They had been lying on the floor for hours before someone came to help them. They did have alarm buttons that they had to hang around their necks, but they were too proud or too lax to wear them. Under pressure, especially from their son Coen, they moved to the Johanna House in Wassenaar, a luxurious nursing home. They had a spacious apartment there, with a private living room, bedroom, bathroom and kitchen. They could not take all their belongings with them. Oort did take most of his books, but a lot was distributed among the children and grandchildren. For example, the sundial in the shape of a radio telescope of Fig. 11.12 went to grandson Marc, with whom it still stands in the garden. They lived there reasonably well on their own, did have good relations with neighbors and other inhabitants of the Johanna House but took little part in the communal activities.

Fig. 12.7 Oort at the reception for his ninetieth birthday, together with three Dutch Directors General of the European Southern Observatory ESO. From left to right: Lo Woltjer (DG 1975–1987), Adriaan Blaauw (1970–1974) and incumbent Harry van der Laan (1988–1992). Provided by the European Southern Observatory

Oort left home less and less after having moved to the Johanna House in Wassenaar. Probably his last participation in a public ceremony was the PhD defense of Rein van de Weygaert (b. 1963) in 1991 (see Fig. 12.8).

The end came in the fall of 1992, when Oort broke his hip in a fall at home. Two days later, on November 5, 1992, Jan Hendrik Oort died from complications. At his bedside were Mrs. Oort, his son Bram and daughter-in-law Bineke. It was somewhat unexpected, his daughter and other son had left for home.

The cremation took place at Westerveld crematorium near Velsen, where his mother had also been cremated. It was on November 10 and was attended by a large number of friends and colleagues alongside the family.

Mrs. Oort lived not longer than another seven months. Johanna Maria (Mieke) Oort-Graadt van Roggen died on June 7, 1993. She also took a bad

Fig. 12.8 Oort's probably last participation in a public ceremony in an academic gown was the PhD defense on September 25, 1991 by M.A.M. (Rien) van de Weygaert. Behind Oort Vincent Icke, who defended his thesis in 1971 in Leiden. Collection Rien van de Weygaert. Collection Rien van de Weygaert

fall and died from the complications. She too was cremated at Westerveld. The famous poem 'The Water Lily' by Frederik Willem van Eeden (1860–1932) appeared on the funeral card [130]:

> I love the immaculate waterlily white,
> so pure, so silent
>   as she unfolds her crown,
> in summer's light.
>
> Rising out of the lightless chilly pond,
> she's found the summer light
> and has gladly unlocked her golden heart.
>
> Now she lies pensive on the water's plane
> and wishes no more...

From the speech of granddaughter Yvette de Smidt at the funeral ceremony for Oort on November 10, 1992 (supplied by Bram Oort):

A very special grandfather, because he took seriously all your chats and stories. Whether you wanted to hoverfly, scuba dive or all those other crazy things we did and do, grandpa reacted stimulatingly as if nothing was too crazy for him, or perhaps rather, because he had a natural confidence in all of us. He asked questions that made us think again, and later asked how the adventures had ended.

We all have been in the dome at the Sterrewacht when we were children and experienced the excitement of the starry sky opening up just for you alone. Later we understood—one maybe more than the other—his passion for the Universe. A year ago he decided to convene his grandchildren so that he could tell of the latest developments. And as everyone knows, when grandpa made a plan it had to be executed. So this spring we experienced his last lecture as a very special happening.

The Haamstede saying: 'O future generations, see this, and remember to stay united', is a wish that has been fulfilled thanks to the strong bond between grandpa and grandma.

Beloved grandpa, thank you for everything.

I close this book with the lines that Bram Oort quoted at the beginning of his contribution to the *Liber Amicorum* in 1980, and that Mrs. Oort and the family quoted (in abbreviated form) on the funeral announcement card of Jan Hendrik Oort. It is by American anthropologist and science writer Loren Eiseley (1907–1977) from his book *The immense journey*.

Down how many roads among the stars must man propel himself in search of the final secret? The journey is difficult, immense, at times impossible, yet that will not deter some of us from attempting it. We cannot know all that has happened in the past, or the reasons for all of these events, any more than we can with surety discern what lies ahead. We have joined the caravan, you might say, at a certain point; we will travel as far as we can, but we cannot in one lifetime see all that we would like to see or learn all that we hunger to know.

Fig. 12.9 Oort on the roof of the Sterrewacht Leiden in 1991. From the Oort Archives

A

Astronomical Background

The universe is full of magical things,
patiently waiting for our wits to sharpen.
Eden Phillpotts (1862–1960).[1]

For many of the things discussed here, particularly concerning the orbits of planets and the structure of stars, a more elaborate treatment using mathematics when helpful, see the online version of the appendix to my Kapteyn biography in this series at www.astro.rug.nl/JCKapteyn/AppendixA.pdf.

A.1 Kepler and the Orbit of Mars

Oort has on many occasions mentioned Jacobus C. Kapteyn as his 'groote leermeester' (great teacher) and inspirer. In an autobiographical article in 1981 he wrote:

> When in 1917, at the age of 17. I began my studies in Groningen I became almost immediately inspired by Kapteyn's lectures on elementary astronomy. [...] Perhaps the most important thing I learned – mainly, I believe from Kapteyn's discussion of Kepler's method of examining nature – was to tie interpretations directly to observations, and to be extremely wary of hypotheses and speculations.

What is this method of Kepler, which Oort was referring to?

It is about Kepler's determination of the orbit of Mars and the discovery of his laws for planetary orbits. It can be found in Oort's notes of Kapteyn's

---

Eden Phillpotts was a British writer and poet.

© The Editor(s) (if applicable) and The Author(s), under exclusive license
© Springer Nature Switzerland AG 2021
P. C. van der Kruit, *Master of Galactic Astronomy: A Biography of Jan Hendrik Oort,*
Springer Biographies, https://doi.org/10.1007/978-3-030-55548-1

lectures, as well as in Oort's notes of his own lectures later on this subject (and as a matter of fact in my own notes to Oort's lecture course on the planetary system in 1964/1965).

Planets describe orbits in the sky relative to the stars. As seen from the Sun, these are large circles in the sky, but from Earth, the orbit across the starry sky is more complicated. In ancient times with the assumption of the Earth being in the center of the universe, these orbits were described as a superposition of two circles, the deferent and the epicycle. For the outer planets Mars, Jupiter and Saturn this looks like Fig. A.1, left. This geocentric lay-out is basically correct, because the description of the position in the sky is about directions and does not use distances. The deferent is then in fact the orbit of the planet and the epicycle that of the Earth (for the inner planets Mercury and Venus it is the other way around). The problem however was that, following Plato (427–347BC) and Aristotle (384–322BC), the motions in the super-lunar, ideal world had to be described with uniform speeds along pure circles. So the deferent and epicycle had to be circles, and the center of the epicycle along the deferent and the planet in the epicycle had to move at constant speed. This was a problem, because—as Kepler would later find in his first two laws— the planets actually move on ellipses with the Sun in one of the focal points and faster as they are closer to the Sun and slower further away. Claudius Ptolemæus (Ptolemy) had 'solved' this by introducing the equant (see Fig. A.1, right), where the Earth is not in the center and the motion in the deferent now is uniform with respect to a point positioned symmetrically opposite from it and not to the center of the deferent.

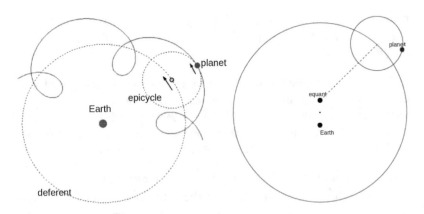

Fig. A.1 The Ancient Worldview (for the outer planets). Left: the deferent is in fact the orbit of the planet and the epicycle that of the Earth. The resulting orbit as seen from the Earth is broadly in line with the observations in terms of directions. Right: the introduction of the equant, in which the center of the epicycle moves with a uniform velocity along the deferent as seen from this equant. Figure by the author

By our standards, that would be cheating, but it is remarkable that this works so well. Seen from the Sun the Earth goes in its orbit at an average angular speed of $0°9856$ per day and this varies over the course of a year with $+3.4$ to -3.4% relative to this. From the point of view of the 'empty' focus, however, this variation is only 0.014%. The Mars orbit is more eccentric than that of the Earth, so that the variation of angular speed seen from the Sun varies from -17.7% relative to the average, to $+19.5\%$. But also in this case, the variation from the empty focus is very small, namely $\pm0.4\%$. This remarkable property of the planetary orbits is exactly what Ptolemy needed for his equant.

Nicolaus Copernicus replaced the geocentric with the heliocentric world-view, but retained the prejudices of Plato and Aristotle. To obtain agreement with observations, he used epicycles again, but now in order to approximate the orbits with what would later become ellipses, a very different reason for introducing epicycles than in antiquity. Kepler had access to the most accurate observations of planets of that time, made by Tycho Brahe (1546–1601). Using these data he was unable to fit the data to such a model. It is this unsatisfactory situation which led Kepler to throw all preconceived notions out of the window. This is what Kapteyn explained in his lectures and this is what inspired Oort to never make more assumptions than is strictly necessary, and to always take observations as a starting point and never preconceived ideas and notions. Kepler asked a question that had not been asked for about two thousand years, namely: What do the *observations* tell us about the orbits of the planets?

According to Oort's notes, the way in which Kepler determined the orbit of Mars (and next also for the other planets) was 'ingenious'. He talked about it with great admiration. How did Kepler actually determined the orbit of Mars?

Kepler had embraced the heliocentric world view and so a planetary orbit was exactly like the deferent in Fig. A.1, on the right, but with the Sun now at the place of the Earth and the epicycle gone. That is important, because if the angular speed from the equant is uniform, that of the planet in the orbit must be variable. So Kepler had already renounced the uniform motion of Plato and Aristotle. For the Earth this was actually known, because the Sun moves irregularly in relation to the stars. The inequality of the length of the seasons was already known in antiquity; after all, to compensate for this February has only 28 (or 29) days (but the same thing could have been accomplished more elegantly by letting December and January have 30 days).

The first step was to determine the period of the orbit of Mars. Kepler had at his disposal many observations of the times when Mars was in opposition, i.e. in the sky directly opposite from the Sun. That means that he could determine in which direction Mars was seen from the Sun at those times. Namely, just the same of that of the Earth from the Sun, which is of course opposite from the direction in which we see the Sun relative to the stars. With that information he then also knew in which direction from the Sun Mars was seen at these times. From a set of such observations he could determine the orbital period of Mars. Kepler found 686.95 days (1.88 years), only 0.03 days shorter than we know now.

Next he determined the *shape* of the orbit of the Earth. For that purpose, he selected observations of Mars at times of which he now knew that Mars had to be in exactly the same position in its orbit (Fig. A.2, left). In the triangle Earth-Sun-Mars he knew the two angles indicated. After all, from the times he knew the directions in which Mars and the Earth were seen from the Sun (and therefore also the angle between them) and also the directions in which the Sun and Mars were seen from the Earth (and thus also the angle between them). Then he could calculate the distance Earth-Sun $r1$ as a fraction of the distance Sun-Mars D. By taking other situations, where Mars is in that same position in its orbit, he could calculate more of such distances Earth-Sun, $r2$, $r3$, etc., all in terms of D. The orbit of the Earth then followed from this, albeit on an unknown scale.

Kepler then took two times for which he knew that Mars was at the same position in its orbit (Fig. A.2, right). So he knew where the Earth was and what the distances $r1$ and $r2$ were. The angle between those directions was known because at those times the direction towards the Sun was known. But he also knew the angle between the directions in which the Sun and Mars were seen from the Earth. That means that the triangle E1-E2-Sun was known and the two directions towards Mars. Where those crossed was then the position of Mars. Kepler repeated this for many such situations and so he found the

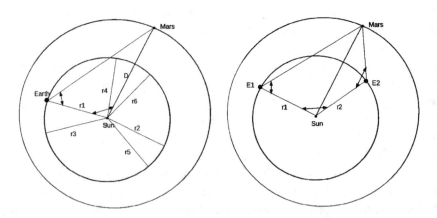

Fig. A.2 Kepler's determination of the orbit of Mars. For details see the text. Figure by the author

orbit of Mars, at least on the same unknown scale. Now that the Mars orbit was known, Kepler turned it around and chose times when the Earth was in the same position and imagined himself to be an observer on Mars. This step required the insight that the direction in which the Earth was seen from Mars was exactly the opposite of the direction in which we see Mars from Earth. In this way, Kepler and mapped all the orbits of the planets of the Solar System. The corresponding scale of this map of the Solar System can then be found by determining at a certain point in time the distance between the Earth and the Sun or the Earth to another object (planet or asteroid).

This method was described by Kapteyn as 'directly linking interpretations to observations, and always being extremely wary of hypotheses and speculations'. The assumption that planetary orbits can only be described with uniform motion in circles has held up progress for many centuries. Oort, in turn, presented the same material in his lectures, in my case in his lecture series 'The Planetary System', which I attended in the academic year 1964/1965. Subsequently, I have done the same in my lectures. All of this has been described in an very fascinating way in: *The Sleepwalkers: A history of man's changing vision of the Universe* by Arthur Koestler [41]. I have given all my Ph.D. students a copy of that book as a present after they had defended their Ph.D. theses.

A.2   Positional Astronomy

Astronomical Coordinate Systems

Positions in the sky are indicated by coordinate systems on a sphere, like geographical longitude and latitude on Earth. The sky is seen as a sphere, the celestial sphere, which appears, as it were, as a large dome over the observer. The simplest coordinates of an object on the celestial sphere (see Fig. A.3, left) are *azimuth*, measured along the horizon usually from the south, and *altitude*, which is the angle above the horizon. However, these change with time and with position on Earth.

For a catalogue we need coordinates which are not dependent on time on the clocks or position on Earth, in other words that are fixed to the sky. The most common ones are *right ascension*, measured along the equator (referred to as R.A. or α), and *declination* (Dec. or δ), perpendicular to it in the direction towards the poles. The poles are the points where the rotation axis of the Earth intersects the celestial sphere. Right ascension is measured from the vernal equinox, the position on the equator where the Sun is located when it crosses the equator around 21 March (Υ in Fig. A.3, right). Right ascension

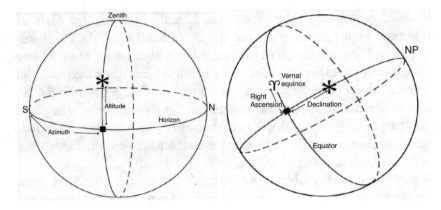

Fig. A.3 Coordinates on the sky of a star in azimuth and altitude (left) and in right ascension and declination (right). The symbol ♈ indicates the vernal equinox. Figure by the author

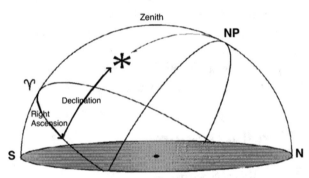

Fig. A.4 The coordinate system of right ascension (R.A.) and declination (Dec.) for an observer in the northern hemisphere. The situation is for case that the vernal equinox is precisely on the meridian (the line south 'S' via the zenith straight up to north 'N') Figure by the author

is measured counterclockwise and is expressed in hours (a full circle of 360° corresponds to 24 h), minutes and seconds (Fig. A.4).

The Sun moves in a year around the celestial sphere along a great circle which is the *ecliptic*. It makes an angle of about 23°5 with the equator and intersects it in the equinoxes, the vernal equinox in March and on the opposite side the autumnal equinox.

Galactic longitude and *latitude* are used for studies of our Galaxy. Longitude is measured along the Milky Way (the plane of the Milky Way System) and latitude is perpendicular to this. The zero point for longitude is the direction towards the center of the Milky Way Galaxy, but for a long time (during Oort's work before the 1950s) it has been the point of intersection of the Milky Way and equator. In the old notation, the center of the Milky Way was at longitude 327°. The Galactic equator (the Milky Way) makes an angle of 62°.9 with the celestial equator.

Precession, Atmospheric Refraction and Aberration

The rotation axis of the Earth makes a slow spinning motion in space around the line perpendicular to the orbital plane, just like a spinning top on a plane surface. This is called *precession*. This results in a change of the positions of the poles and therefore also of the equinoxes over time. The poles complete a full circle along the equator in a period of about 26,000 years. This corresponds to a motion of the equinoxes along the celestial equator of about 50 seconds of arc per year. The R.A. and Dec. of each object also change as a result of this. The phenomenon was discovered around 130BCE by Hipparchus of Nicaea (ca.190–ca.120BCE). The slope of the ecliptic (the annual orbit of the Sun on the sky) relative to the equator also changes slowly due to disturbances of the Moon and other planets on Earth. This is called *nutation*.

The Earth's atmosphere is curved just like its surface, and its density changes with altitude. The light of a star that passes through the atmosphere is refracted and the star as a result seems to be slightly higher above the horizon than if there were no atmosphere. This is called atmospheric refraction. This effect depends on the wavelength of the light (it is stronger in blue than in red light) and at average optical wavelength (green light) at a height above the horizon of 45° it amounts to about a minute of arc.

Aberration of light is due to the finite speed of the light (300,000 km/s). The effect is usually compared to the sloping tracks of raindrops along the window of a moving train. It is the result of the fact that the telescope has been moved slightly by the motion of the Earth around the Sun (about 30 km/s) during the time the light travels through the telescope tube. It depends on the direction relative to the Earth's motion and is up to 30/300,000 rad or about 20 arc seconds. It has been discovered by James Bradley (1693–1762) in 1725.

A.3 Properties of Stars

Distances of Stars

In the course of a year, a star describes a small elliptical orbit on the sky as a reflection of the orbital motion of the Earth around the Sun (see Fig. A.5). The semi-major axis is called the *parallax*. This was for the first time measured in the 1830s at the Königsberg Sternwarte (now Kaliningrad, Russia) by Friedrich Wilhelm Bessel (1784–1846) for the star 61 Cygni, a by then already well-established binary star, at the Dorpat Observatory (now Tartu) in Estonia

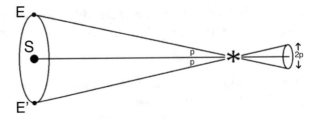

Fig. A.5 The parallax of stars. The motion of the Earth in its orbit around the Sun results in a motion of a star along a small ellipse on the sky. The size of the ellipse is smaller when the star is further away from us. Figure by the author

by Friedrich Georg Wilhelm von Struve (1793–1864) for Vega, one of the brightest stars in the northern hemisphere, and at the Royal Observatory at the Cape by Thomas James Henderson (1798–1844) for the nearest star α Centauri.

Distance of stars are expressed in *parsec*, which is the distance of a star whose parallax is one second of arc. Herbert Hall Turner (1861–1930), a British astronomer, is considered to have been the person who introduced it. A parsec is 3.1×10^{16} m or about 206,000 Astronomical Units (the average distance Earth-Sun). This is equal to the number of arcseconds in a radian. Sometimes the *lightyear* is used, the distance light travels in a year; one parsec is equal to 3.26 lightyears.

Brightness of Stars

For historical reasons astronomers express the apparent brightness of stars in *(apparent) magnitudes*. This concept originates from antiquity, categorizing the brightest stars as being of magnitude one and the faintest one could see as magnitude six. Eventually, it has been defined precisely as a logarithmic scale, in which 5 magnitudes corresponds to a factor of 100 (one magnitude is then a factor 2.512) in brightness. This was proposed by British astronomer Norman Robert Pogson (1829–1891), who introduced the scale in 1856.

A bright star like Sirius has a magnitude −1.5; the faintest stars visible to the naked eye magnitude 6, which corresponds to a factor of 1000 or so in brightness. The Pole Star (Polaris) has magnitude 2.0, and the majority of the stars in the well-known constellations Ursa Major (Big Dipper) and Cassiopeia or in the 'belt' in Orion (accent on the second syllable 'ri') have a magnitude between 1.7 and 2.4, which is a variation of only a factor two or so among them. But the distances of these stars vary between 65 and 1300 lightyears, or a factor of 20.

To correct the apparent magnitude of a star for its distance, astronomer use the *absolute magnitude*, which is defined as the apparent magnitude, that the star would have if the distance were 10 pc. This definition originates from Kapteyn (see my biography of Kapteyn for details and background). If m is the apparent magnitude and M is the absolute magnitude, and r the distance in parsec (or p is the parallax in arcseconds), we have

$$M = m + 5 - 5\log r = m + 5 + 5\log p.$$

For the Sun, the absolute magnitude is $M = 4.8$ at visual wavelengths.

Motions of Stars

Over time, the position of a star in the sky changes due to its motion in space relative to us. This is called the *proper motion*. It was first noticed by Edmond Halley in 1718.

In addition to the motion of stars as a result of their random motion through space, there is also a systematic pattern across the sky, which results from the motion of the Sun itself relative to the stars in its neighborhood. This pattern can be seen as a tendency for the proper motion of stars to point on average away from the Solar Apex. This is the direction towards which the Sun moves and is located in the constellation Hercules. The speed of the Sun relative to the stars around it is about 20 km/s. This flow pattern can be used to statistically determine the distances of a well-defined group of stars, as Kapteyn did. This is called the *secular parallax*, and it works as follows.

Suppose the Sun moves through space at a velocity $V_\odot$. If a star in the sky is an angle λ from the Apex, the component of the solar velocity that is perpendicular to the line of sight $T_\odot = V_\odot\sin\lambda$. The expected proper motion $v_\odot$ resulting from the Sun is along an arc from the star to the Apex and has a value of $v_\odot = pT_\odot/4.74$, where p is the parallax of the star in arcseconds and $r = 1/p$ is the distance in parsecs); the number 4.74 serves to convert the units kilometer per second of T into arcseconds per year for v.

For a group of stars, the secular parallax is the average value of p, as estimated with this relation. In addition, each star receives a weight corresponding to how much the speed of the Sun contributes to the proper motion of that star. This weight is $\sin\lambda$. The secular parallax of a group of stars then is

$$\langle p \rangle = \frac{4.74\langle v\sin\lambda \rangle}{V_\odot\langle \sin^2\lambda \rangle},$$

where the pointed brackets stand for averages.

The *tangential velocity* (T) is the component of the star's velocity perpendicular to the line of sight. If v is the proper motion, corrected for the Solar motion, then:

$$T = 4.74\frac{v}{p} = 4.74vr.$$

The *radial velocity* is the component along the line of sight. It can be measured in spectra with the Doppler effect, named after Christian Andreas Doppler (1803–1853). It is the well-known phenomenon that the tone of an ambulance coming towards us has a higher pitch than when it moves away from us. For light this also holds true and when the speed of a star points away from us, the wavelength increases and the color becomes redder. The change in wavelength compared to the wavelength λ at zero speed equals the radial velocity V_{rad} divided by that of light c:

$$\frac{\Delta\lambda}{\lambda} = \frac{V_{rad}}{c}.$$

This applies as long as the radial velocity is much smaller than that of light. At optical wavelengths—say 5000Å or 500 manometers—a shift of 1 Å corresponds to a radial velocity of about 60 km/s, since c is 300,000 km/s.

A.4   Stellar Evolution

The following is taken from my scientific biography of J. C. Kapteyn: *Jacobus Cornelius Kapteyn, The born investigator of the heavens* [3], with a few editorial changes.

The fundamental diagram in astrophysics is named after the Danish astronomer Ejnar Hertzsprung (1873–1967) and the American Henri Norris Russell (1877–1957) (see Fig. A.6). Hertzsprung worked at that time at the Potsdam Astrophysikalisches Observatorium and was married to Kapteyn's second daughter. Later, Hertzsprung became professor of astrophysics in Leiden and eventually director of the Observatory.

It is a diagram that plots the surface temperature of stars against the luminosity. The vertical axis is the luminosity, but the absolute magnitude can be used for this as well. Along the horizontal axis, which relates to the surface temperature of the star we may also use the color index, for example the difference between the magnitudes of the star in a blue ('B') and a visual ('V') wavelength band. If the star is relatively bright in blue, then the color index (B − V) is relatively small (or negative). The star then is relatively hot at the surface. Alternatively, horizontally we may use the spectral type. The absorption lines in the spectrum of the star are formed by atoms or ions, which absorb

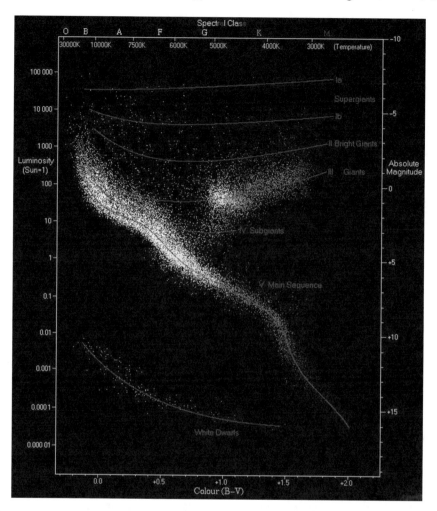

ig. A.6 Hertzsprung–Russell diagram of stars. Horizontally, at the bottom the color ndex $(B - V)$ and along the top the spectral type and the surface temperature in Kelvin. he vertical axis has on the right the absolute magnitude and on the left the luminosity xpressed in that of the Sun. The lines and Roman numerals indicate brightness classes. ee also the text. From *An Atlas of the Universe* [131]

ight at specific wavelengths, and which atoms or ions are present and which ines are prominent is strongly influenced by the temperature.

Spectral types are indicated by the letters $OBAFGKM$ and a decimal ubdivision within them, based on precisely defined spectral line strength ratios. y definition, the color index $(B - V)$ is zero for an $A0$ star. O-stars are elatively blue and along the sequence of spectral types, the color becomes ncreasingly red and the temperature at the surface of the star lower. In O-stars the lines of ionized helium are strongest, in B-stars those of neutral helium, in A-stars of hydrogen, while in F, G and K-stars ionized calcium and lines of metals gradually become stronger. In M-stars, lines of molecules (such as titanium oxide) are strong.

The radiation of a star is very close to so-called black-body radiation. Each body with a temperature T emits an amount of radiation integrated over all wavelengths, which depends on the temperature T. According to Stefan–Boltzmann's law, this corresponds to an energy per second per unit of surface area, which is equal to σT^4 (with the constant σ also named after these two physicists). And since the surface of a sphere with radius R is equal to $4\pi R^2$, it follows that the luminosity L of a star is:

$$L = 4\pi \sigma R^2 T^4.$$

Most stars lie along the Main Sequence, which runs from top left to bottom right. After having formed by the contraction of a cloud of gas, which releases potential energy, stars become hotter in their central volumes, because during the contraction more potential energy is freed in the form of heat than can be radiated away. At ten million degrees, nuclear reactions start to occur, converting hydrogen into helium. O-stars at the top of the Main Sequence are bright, heavy and hot, and are also remain in this stage for the shortest period. M-stars are weak, light and cool and live longer than the present age of the universe. If L is the luminosity, T the temperature at the surface, M the mass and t the time stars remain on the Main Sequence, then we have in a reasonable approximation:

$$L \propto T^6 \quad L \propto M^3 \quad \tau \propto M^{-2}.$$

In time, all the hydrogen in the central parts will have been used up and that part will stop producing energy. Initially, 'hydrogen burning' will continue for some time in a shell around the core and the star will become brighter. A star like the Sun then moves along the sub- and the giant branch (along numbers IV and III in Fig. A.6). The temperature at the surface drops and the star becomes redder, but according to the formula above, the radius R must then increase. Because energy is no longer produced in the central parts, that part contracts but then becomes hotter again. When this core becomes sufficiently hot (of the order of a hundred million K), the helium will start 'burning' there, which then is converted into carbon and oxygen. A star like the Sun is then located on the 'clump' halfway along the line III. When the helium in the central parts is also exhausted, the inner parts of the star will collapse further and the star will repel its outer layers and become a so-called planetary nebula. The remains of the core cool down to become a white dwarf on the line at the bottom of the figure. Matter becomes very compact, but at a certain point the contraction of will stop because Fermi's exclusion principle and Heisenberg's uncertainty relationship together forbid the electrons to come even closer together (this is so-called degeneration pressure).

In a more massive star, however, the pressure is so great that the core will continue to contract and it then becomes so hot that even heavier chemical elements are formed. But after a relatively short period of time, this also stops and the central parts pull together even further. The contraction force is then so great that even the degeneration pressure can no longer compensate it and then the electrons are, as it were, pressed into the protons and they form neutrons. The process is so fast that an enormous pressure wave passes through the star, blowing it up as a supernova. The chemical elements formed in the course of evolution are added to the interstellar gas. But also more chemical elements are formed and this matter is also thrown into space. The remaining central part is then a neutron star that is stable because of the neutron's degeneration pressure, or, if the star mass is even larger, the contraction forces are even too strong for that, and the star becomes a black hole.

The derivation of the formulae above and more background on the issues in this section can be found among others in my introductory astronomy lectures [132].

```
                    JAN  HENDRIK  OORT
                    -------------------

                    Curriculum Vitae


Born: Franeker, the Netherlands, on April 28, 1900.
Married: to Johanna Maria Graadt van Roggen, May 24, 1927.
Three children.


Secondary school Leyden (certificate 1917).
Student at the Groningen University from 1917 to 1921 (Professors of astronomy:
          J.C. Kapteyn and P.J. van Rhijn).
Doctor of astronomy at Groningen 1926.

1921-'22  Research assistant at the Laboratory of Astronomy at Groningen.
1922-'24     "       "      "   "  Yale Observatory
Since 1924 Astronomer at the Leyden Observatory
1926-'30   Lecturer at the Leyden University.
1930-'35   Assistant-professor at the Leyden University.
1935-'45   Associate professor at the Leyden University and
           Associate director of the Observatory.
1945-'70   Professor and Director of the Leyden Observatory.
1949-'70   President of the Netherlands Foundation for Radio Astronomy.
1952       "Visiting professor" at the California Institute of Technology
           at Pasadena and at the University of Princeton.
1956       Waynflete Reader, Magdalen College, Oxford.
1958       Special Fellow Carnegie Institute and Alexander Morrison Research
           Associate, University of California.
1935-'48   General Secretary of the International Astronomical Union.
1958-'61   President of the same Union.


Honorary degrees: Copenhagen, Glasgow, Oxford, Louvain, Harvard, Brussels,
                  Cambridge, Bordeaux, Canberra, Torún.
Gold Medal of the Royal Astronomical Society, London.
Bruce Gold Medal of the Astronomical Society of the Pacific, San Francisco.
Janssen Medal of the Société Astronomique de France, Paris.
Janssen Prize of the Société Astronomique de France, Paris.
Vetlesen Prize of Columbia University of New York.
Prize of the International Balzan Foundation, Milan.

Correspondent of the Bureau des Longitudes, Paris.
Member of the Royal Academy of Sciences at Amsterdam.
Foreign Associate of the Academy of Sciences at Paris.
Honorary member of the Royal Institution, England.
Foreign member of the Royal Society, England.
Foreign Associate of the National Academy of Sciences, Washington.
Foreign member of the Academy of Sciences, Moscow.
Member of the Pontifical Academy of Sciences.
Ridder in de Orde van de Nederlandse Leeuw.
Commandeur in de Orde van Oranje Nassau.
```

Fig. B.0 Oorts curriculum vitae as he drew it up himself in the mid-1980s. From the Oort archives

B

Honors and Genealogy

Honor to whom honor is due.
Dutch proverb.

B.1 Awards, Prizes and Honorary Doctorates

Figure B.0 shows a reproduction of Oort's curriculum vitae, as drawn up by himself. Figures B.1 and B.2 show a list of his honors and distinctions, again drawn up by himself and in his own handwriting. The final one is the Balzan Prize for Astrophysics in 1984. There are two more to be added: the Kyoto Prize for Astronomy in 1987 and an emeritus membership of the Academia Europaea in 1989.

The list with details can be found in the inventory of the Oort Archives in Jet Katgert-Merkelijn's *JKM-inventory* [7], item 333: 'Memberships of Academies and Learned Societies, Honors, Awards and Prizes (1935–1989)', starting on p. 149. The winning essay with Jan Schilt for the competition of the Bachiene Foundation in 1923 is missing from that list.

There are 22 memberships of academies and other learned societies in the Netherlands, France, the United Kingdom, the United States, Belgium, Sweden, Portugal, South Africa, Canada, Germany, Italy, the Vatican and the Soviet Union. Oort received ten honorary doctorates, namely from the Universities of Copenhagen, Denmark (1946), Glasgow, United Kingdom (1950), Oxford, United Kingdom (1951), Leuven, Belgium (1955), Harvard, United States

© The Editor(s) (if applicable) and The Author(s), under exclusive license
© Springer Nature Switzerland AG 2021
C. van der Kruit, *Master of Galactic Astronomy: A Biography of Jan Hendrik Oort*,
Springer Biographies, https://doi.org/10.1007/978-3-030-55548-1

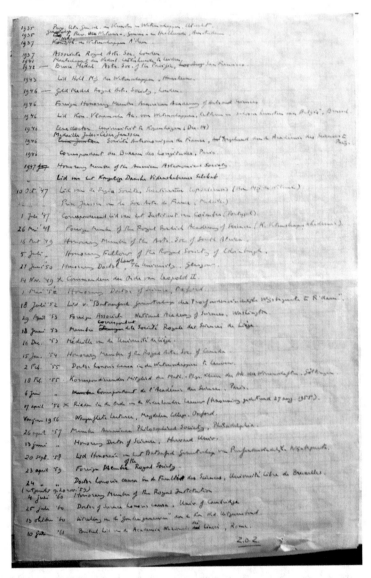

Fig. B.1 Oort's handwritten list of prizes, awards and other honors (continued in Fig. B.2). From the Oort archives

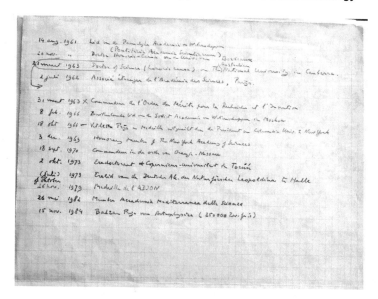

Fig. B.2 Oort's handwritten list of prizes, awards and other honors (continued from Fig. B.1). It ends with the Balzan Prize (1984); the Kyoto Prize for Astronomy (1987) and an emeritus membership of the Academia Europaea (1989) are missing. From the Oort archives

1957), Brussels, Belgium (1959), Cambridge, United Kingdom (1960), Bordeaux, France (1961), Canberra, Australia (1963) and Torún, Poland (1973).

The most important prizes, medals and lectures are the Bruce Medal of the Astronomical Society of the Pacific (1942), the Gold Medal of the Royal Astronomical Society (1946), the Jules-César Janssen Medal of the Société Astronomique de France (1946), the Medal of the Université de Liege (1953), the Waynflete Lectures at Magdalen College, Oxford (1956), the 'Gouden Ganzeveer' (Golden Quill) of the Royal Dutch Publishers' Union (1960), The Vetlesen Prize of Columbia University (1966), the Medal de l'ADION, Observatoire de la Côte d'Azur (1979), the Balzan Prize for Astrophysics (1984) and the Kyoto Prize for Astronomy (1987).

Royal and other high distinctions are Commander of the Order of Leopold I (1949), Knight of the Order of the Dutch Lion (1956), Commander of the Ordre de Merité pour la Recherche et l'Invention (1963) and Commander of the Order of Orange-Nassau (1970).

Asteroid 1691 Oort was discovered in 1956 in Heidelberg by Ingrid Groeneveld (who was married the year before to Kees van Houten) and local staff member K. Reinmuth. When the elements of the orbit had been determined sufficiently well, they were allowed to propose a name to the International Astronomical Union IAU. It is a somewhat irregular rock with dimensions of

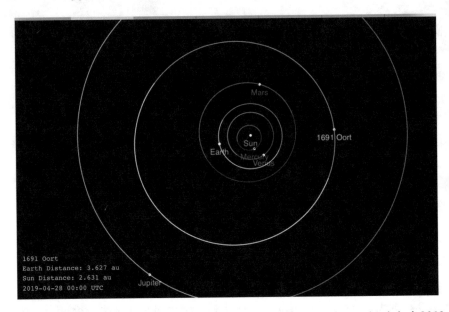

Fig. B.3 Asteroid 1691 Oort. The position is that for 28 April (Oorts birthday) 2019. Produced with the JPL Small-Body Database Browser [133]

30 to 35 km and rotates around its axis in 10.3 h. It goes around the Sun in 5.62 years in a somewhat elliptical orbit with a semi-major axis of 3.2 Astronomical Units. The orbital plane slopes by about 4 degrees relative to that of the Earth. The position together with that of the planets up to Jupiter on 28 April 2019 (Oort birthday in 2019) is shown in Fig. B.3.

Figures B.4 and B.5 show two public works of art dedicated to Oort.

B.2   Academic Genealogy

Below is an abridged version of Oort's academic genealogy. A more extensive version with references and links can be found on my homepage (because as a student of Oort it is also my genealogy): www.astro.rug.nl/~vdkruit.

In principle, it involves academic Ph.D. dissertations, where a doctoral degree has been obtained with the previous person as supervisor. However, this is not always possible. I presented Johann Samuel König as a student of the Bernoullis and von Wolff, although there was no formal Ph.D. thesis defense. Gerrit Moll's doctorate honoris causa was also allowed. I start with Oort's supervisor van Rhijn.

Fig. B.4 In Leiden, a number of 'Wall formulas' have been painted on houses. They are all connected in some way to Leiden University [134]. The house, on which the Oort constants have been painted, is located to the west of the Observatory just on the other side of the canal, the Witte Singel. The lower part of the figure shows the formulas themselves. See also [135]. Photograph by the author

Fig. B.5 The 'Oort Cloud', part of the '11Fountains project', in which a fountain wa built in each of the eleven Frisian cities. It is located in Franeker, the city where Oor was born. It was designed by the French artist Jean-Michel Othoniel (b. 1964) [136]. Th golden, open grid disperses a mist of water droplets, depicting the Oort Cloud. Mad available by 11fountains.nl, reproduced with permission

Pieter Johannes van Rhijn (1886–1960)—student of J. C. Kapteyn: Ph.D. thesis: *Derivation of the change of colour with distance and apparent magnitude together with a new determination of the mean parallaxes of the stars with given magnitude and proper motion,* Groningen, 1915.

Jacobus Cornelius Kapteyn (1851–1922)—student of C. H. C. Grinwis: Ph.D. thesis: *Investigation of vibrating flat membranes,* Utrecht, 1875.

Cornelis Hubertus Carolus Grinwis (1831–1899)—student of R. van Rees: Ph.D. thesis: *De distributione fluidi electrici in superficie conductoris* (on the distribution of electricity over the surface of a conductor), Utrecht, 1858.

Richard van Rees (1797–1875)—student of G. Moll: Ph.D. thesis: *De celeritate soni per fluida elastica propagati* (about the speed of sound in an elastic fluid), Utrecht, 1819.

Gerrit (Gerard) Moll (1785–1838)—student of J. T. Rossijn: Doctorate honoris causa, Utrecht, 1815.

Johannes Theodorus Rossijn (1744–1817)—student of A. Brugmans: Ph.D. thesis: *De tonitru et fulmine ex nova electricitatis theoria deducendis* (on thunder and lightning according to the new theory of electricity), Franeker, 1762.

Antonius Brugmans (1732–1789)—student of J. S. König: Ph.D. thesis: *Disertatio philosophica inauguralis de phaenomeno* (philosophical dissertation on the phenomena), Franeker, 1749.

Johann Samuel König (1712–1757)—student of Johann and Daniel Bernoulli and Christian Wolff. König studied in Basel under Johann Bernoulli from 1730 and under Daniel Bernoulli from 1733. He did not obtain a formal doctoral degree. In 1735 König went to Marburg to study with Christian von Wolff.

Daniel Bernoulli (1700–1782)—student of Johann Bernoulli(?): Ph.D. thesis: *Dissertatio physico-medica de respiratione* (dissertation on the medical physics of breathing), Basel, 1721.

Johann Bernoulli (1667–1748)—student of Jacob Bernoulli: Ph.D. thesis: *Dissertatio physico-anatomica de motu musculorum* (dissertation on the physics and anatomy of muscle movement), Basel, 1694.

Jacob Bernoulli (1654–1705)—student of G. W. von Leibniz: Ph.D. thesis: *Solutionem tergemini problematis arithmetici, geometrici et astronomici* (solutions to a triple problem in arithmetic, mathematics and astronomy), Basel, 1684.

Christian von Wolff (1679–1754)—student of E. W. von Tschirnhaus and G. W. von Leibniz: Ph.D. thesis: *Dissertatio Algebraica de Algorithmo Infinitesimali Differentiali* (dissertation on the algebra of solving differential equations using infinitesimals), Leipzig, 1704.

Ehrenfriend Walter von Tschirnhaus (1651–1708)—He studied in Leiden from about 1669 to 1674, but there is no indication that he obtained a doctorate.

Gottfried Wilhelm von Leibniz (1646–1716)—student of E. Weigel: Ph.D. thesis: *De casibus perplexis in jure* (about disconcerting cases in law), Althof, 1666.

Erhard Weigel (1625–1699)—student of P. Müller: Ph.D. thesis: *De ascensionibus et descensionibus astronomicis dissertatio* (astronomical dissertation on rising and setting), Leipzig, 1650.

Philipp Müller (1585–1659)—He was the supervisor of Erhard Weigel and professor of mathematics at the University of Leipzig from 1616. Müller had a keen interest in astronomy and was one of the first to adopt Kepler's laws of planetary orbits. He corresponded extensively with **Johannes Kepler** (1571–1630). It is known that Jacob Bartsch (ca.1600–1633), a pupil of Philip Müller made a star chart based on data from Müller's data and later became Kepler's assistant and son-in-law.

Oort was probably not familiar with this genealogy. But it would undoubtedly have given him satisfaction (as it does me as a student of Oort) that it goes back to someone that close to Kepler.

Fig. C.0 Oort in his study in the Huygens building on the Wassenaarseweg in Leiden, where the Sterrewacht has been housed since 1973. From the Oort archives

C

Literature

Do I ever get tired of writing?
No, I can do it sitting down.
Simon Vestdijk (1898–1971).[2]

References are given below that are indicated in the text. Only the first 8 are general references to material about Oort and his teacher Jacobus C. Kapteyn. These are arranged chronologically. My Oort Website www.astro. rug.nl/JHOort contains a complete list of references to publications and other material from and about Oort.

References to astronomical articles are given to the NASA Astronomy Data System ADS, ui.adsabs.harvard.edu/. These are codes in square brackets, such as [1922ApJ....55.. 302K]. Then ui.adsabs.harvard.edu/abs/ 1922ApJ....55.. 302K gives a link to that article. The ADS has scanned the astronomical literature as completely as possible. It provides links to electronic versions of many, particularly older, publications, in the form of .pdf or .gif format format). For more recent articles, 'open access' links can be found where possible.

No Dutch or Flemish author ever won the Nobel Prize for literature. Possibly Vestdijk came closest. between 1958 and 1966 (the last year for which the Nobel Archives are public) Vestdijk was nominated fteen times.

© The Editor(s) (if applicable) and The Author(s), under exclusive license
© Springer Nature Switzerland AG 2021
C. van der Kruit, *Master of Galactic Astronomy: A Biography of Jan Hendrik Oort*,
Springer Biographies, https://doi.org/10.1007/978-3-030-55548-1

Below notes with references have been listed that appear in the text. They are arranged by chapter but numbered sequentially. Where possible, links to electronic publications are provided. URLs that would run into the right margin have been broken without hyphenation. *The notes below contain only references and no clarifications or backgrounds to the text of the book.*

Oort in the garden of his home at the Sterrewacht in 1969. From the Oort archives

References

1. *Oort and the Universe. A sketch of Oort's research and person*, by H. van Woerden, W.N. Brouw, & H.C. van de Hulst, Reidel, ISBN-10: 9-027-71180-1, (1980).
2. *Jan Hendrik Oort: Master of the Galactic System*, by P.C. van der Kruit, Astrophysics and Space Science Library, Vol. 459, Springer Publishers, ISBN 978-3-319-10875-9 (2019), www.astro.rug.nl/JHOort.
3. *Jacobus Cornelius Kapteyn: Born investigator of the Heavens*, by P.C. van der Kruit, Astrophysics and Space Science Library, Vol. 416, Springer Publishers, ISBN 978-3-319-10875-9 (2015), www.astro.rug.nl/JCKapteyn.
4. *De inrichting van de hemel: Een biografie van astronoom Jacobus C. Kapteyn*, by P.C. van der Kruit, Amsterdam University Press (2016), ISBN 978-94-6298-042-6. English translation *Pioneer of Galactic Astronomy: A biography of Jacobus Cornelius Kapteyn* Springer Biographies, Springer Publishers, ISBN 978-3-030-55422-4 (2021).
5. *Horizonnen: Een biografie van astronoom Jan Hendrik Oort*, by P.C. van der Kruit, Prometheus, ISBN 978-90-446-4144-8 (2019).
6. Interview by David DeVorkin of J.H. Oort, November 10, 977, American Institute of Physics, www.aip.org/history-programs/niels-bohr-library/oral-histories/4806.
7. *The letters and papers of Jan Hendrik Oort*, by J.K. Katgert-Merkelijn, Astrophysics and Space Science Library Vol. 213, Kluwer, ISBN: 0-79234-542-8, (1997).
8. *Jan Oort, Astronomer*, by J.K. Katgert-Merkelijn & J. Damen, catalogue of the exhibition on the occasion of Oort's hundredth birthday in 2000,

© The Editor(s) (if applicable) and The Author(s), under exclusive license
ⓘ Springer Nature Switzerland AG 2021
: C. van der Kruit, *Master of Galactic Astronomy: A Biography of Jan Hendrik Oort*,
 pringer Biographies, https://doi.org/10.1007/978-3-030-55548-1

Leiden University Library, ISSN 0921-9293, vol. 35 (2000), openaccess.leidenuniv.nl/handle/1887/77628.

9. *The rise of radio astronomy in the Netherlands: The people and the politics*, by A. Elbers, Springer, ISBN 978-3-319-49079-3 (2017).

10. *Early Dutch radio astronomy (1940-1970): The people and the politics*, Universiteit Leiden (2015), online: openaccess.leidenuniv.nl/handle/1887/36547.

Preface

11. *The legacy of J.C. Kapteyn: Studies on Kapteyn and the development of modern astronomy*, by P.C. van der Kruit & K. van Berkel, Springer, ISBN 0-7923-6393-0 (2000).

12. *Ancestry*, www.ancestry.com/genealogy.

13. *Pondes*, www.pondes.nl.

14. *Geni*: www.geni.com.

15. *Online Familieberichten*, www.online-familieberichten.nl.

16. *Stamboomzoeker*, www.stamboomzoeker.nl.

17. *GenVer*, www.genver.nl.

18. *Delpher*, www.delpher.nl.

19. *Internationaal Instituut voor Sociale Geschiedenis*, www.iisg.nl/hpw/calculate.php.

20. *Currencyconverter*, www.historicalstatistics.org/Currencyconverter.html.

21. *Dutch Ancestry Coach*, www.dutchancestrycoach.com/content/how-much-did-you-say-converting-dutch-historic-currencies (discontinued).

1. Youth in Oegstgeest and Leiden

22. www.clker.com/clipart-kaart-nederland-jan.html; public domain clip art.

23. *Veertigjarig Artsjubileum Dr. A.H. Oort*, Nederlands Tijdschrift voor de Geneeskunde, 80, 3207-3208 (1936). www.ntvg.nl/system/files/publications/1936132070002a.pdf.

24. www.familyberry.com/?q=87A077CFB3B70E5726C535F305B0BC2C.

25. allefriezen.nl/zoeken/persons?f={"search_s_register_gemeente": {"v":"Franeker"}}&ss={"q":"Oort"}page&=3.

26. *Jaarboekje 1904 - Historische vereniging Oud Leiden* www.oudleiden.nl/pdf/1904.pdf.

27. upload.wikimedia.org/wikipedia/commons/thumb/c/c3/Exterieur_overzicht_villa,_behorend_bij_Jelgersmakliniek_-_Oegstgeest_-_20310532_-_RCE.jpg.

28. *De tweede Gouden Eeuw: Nederland en de Nobelprijzen voor natuurwetenschappen 1870-1940*, by B. Willink, ISBN10 9-035-11942-8, Bakker (1998). www.dbnl.org/tekst/tekst/will078twee01_01/.

29. *Origins of the Second Golden Age of Dutch Science after 1860*, by B. Willink. sss.sagepub.com/content/21/3/503.abstract.
30. *Heike Kamerlingh Onnes: de man van het absolute nulpunt*, by D. van Delft, ISBN10 9-035-12917-2, ISBN13 9-789-035129-177, Bakker (2005). Electronic version available at www.dbnl.org/tekst/delf006heik01_01/.
31. www.erfgoedleiden.nl/ae887bbc-26bc-11e3-bc3a-3cd92befe4f8.
32. *Levensbericht van Arend Joan Petrus Oort (1903–1987)*, Royal Netherlands Academy of Sciences, www.dwc.knaw.nl/DL/levensberichten/PE00002151.pdf.
33. www.erfgoedleiden.nl/ca8b3656-26bc-11e3-8138-3cd92befe4f8.
34. Amersfoorts Dagblad/de Eemlander, June 30, 1932, archiefeemland.courant/issue/ADDE/1932-06-30/edition/0/page/6.

2. Kapteyn and Galactic Astronomy Around 1920

35. *De ontdekkers van de hemel*, by D. Baneke, Prometheus, ISBN10 9-035-13688-8 (2015).
36. local.strw.leidenuniv.nl/album1908/book_info.html.
37. *J.C. Kapteyn; Zijn leven en werken*, by H. Hertzsprung-Kapteyn, Wolters 1928.
38. commons.wikimedia.org/wiki/File:PJvanRhijn.jpg.
39. *The motions of stars in a Kapteyn Universe*, by J.H. Jeans, Monthly Notices of the Royal Astronomical Society, 82, 122-132 (1922). [ADS: 1922MN-RAS..82..122J].

3. Student at the University of Groningen

40. *Some notes on my life as an astronomer*, by J.H. Oort, Annual Review of Astronomy & Astrophysics, 19, 1-5 1981. [ADS: 1981ARA&A..19....1O].
41. *The sleepwalkers: A history of man's changing vision of the Universe*, by A. Koestler, ISBN13: 978-01401-9246-9 (1959).
42. *Iets over het werk van Prof. J.C. Kapteyn*, by J.H. Oort, Groningsche Studentenalmanak voor het jaar 1922, vier-en-negentigste jaargang, 202-215 (1921).
43. *My first 72 years of astronomical research: Reminiscences of an astronomical curmudgeon, revealing the presence of human nature in science*, privately published by Willem J. Luyten, 1940 East River Road, Minneapolis, MN 55414, USA. 20+203 pp. (1987).

4. Via Yale to Leiden

44. From *No parking. no halt. success non stop!* by A. Ratna. Supernova Publishers (2015), ISBN 8-189-93098-2.
45. *Honderd jaar Leidse Sterrewacht*, Sterrewacht Leiden (1965).
46. *De Leidse Sterrewacht: Vier eeuwen wacht bij dag en bij nacht*, by G. van Herk, H. Kleibrink, & W. Bijleveld, Waanders/De Kler, Zwolle (1983).
47. *De Leidse Sterrewacht: Glorieus als vanouds*, by F.P Israel and the University of Leiden (2011).
48. *Een passie voor precisie: Frederik Kaiser en het instrumentarium van de Leidse Sterrewacht*, by H. Hooijmaijers, Studium, 4, 195-126 (2011). www.gewina-studium.nl/index.php/studium/article/view/1545/7241.
49. Susanne Elisabeth Nørskov, AU Library, Fysik & Steno, Institut for Fysik og Astronomi, Aarhus Universitet, Denmark.
50. *Als bij toverslag, De reorganisatie van de Leidse Sterrewacht, 1918-1924* by D.M. Baneke, BMGN/Low Countries Historical Review 120, 207-225 (2005).
51. *Hij kan toch moeilijk de sterren in de war schoppen. De afwijzing van Pannekoek als adjunct-directeur van de Leidse Sterrewacht*, by D.M. Baneke, Gewina 27, 1-13 (2004).
52. *The interallied conference of scientific academies in London*, by A.G. Marshall, Publications of the Astronomical Society of the Pacific, 30, 331-335 (1918). [ADS: 1918PASP...30..331M].
53. *History of the IAU, Birth and first half-century of the International Astronomical Union*, by A. Blaauw, Kluwer (1994), ISBN 0-7923-2979-1.
54. *The International Astronomical Union: Uniting the community for 100 years*, by J. Andersen, D.M. Baneke, & C. Madsen, Springer, ISBN 978-3-319-96964-0 (2019).
55. www.astro.rug.nl/JCKapteyn/Statement_USAcad.pdf.
56. commons.wikimedia.org/wiki/Category:Rotterdam_(ship,_1908)?uselang=nl#media/File:Rotterdam-Ship.jpg.
57. Yale University buildings and grounds photographs, 1716-2004 (inclusive). Manuscripts & Archives, Yale University, images.library.yale.edu/madid/.
58. University of Chicago Photographic Archive, apf6-04494, Special Collections Research Center, University of Chicago Library.
59. had.aas.org/resources/aashistory/early-meetings/1922-1927#31.

5. Rotation and Dynamics of the Milky Way Galaxy

60. www.universetoday.com/30710/galaxy-rotation/.
61. commons.wikimedia.org/wiki/File:NIMH_-_2011_-_0300_-_Aerial_photograph_of_Leiden,_The_Netherlands_-_1920_-_1940.jpg.

62. *Kosmos*, by W. de Sitter, Harvard University Press, ISBN 97-806-743314-71 (1932).

63. *Niet-lichtgevende materie in het sterrenstelsel*, by J.H. Oort, Inaugural lecture as privaat-docent, Hemel & Dampkring 25, 13-21 + 60-70 (1927).

6. The Structure of Our and Other Galaxies

64. *The intelligent man's guide to the sciences, Vol. I, The physical sciences*, by I. Asimov, Basic Books, New York (1960).

65. *My sister's keeper*, by J.L. Picoult, Atria, ISBN: 0-7434-5452-9 (2004).

66. *Short history of the Observatory of the University at Leiden 1633–1933, published at the occasion of the celebration of the 300th anniversary of the foundation of the Observatory*, by W. de Sitter, Enschedé en Zonen, Haarlem (1933).

67. A. Zezas, J. Huchra, K. Kuntz, F. Bresolin, J. Trauger, J. Mould, Y.-H. Chu & Davide De Martin, heritage.stsci.edu/, hubblesite.org/images/gallery.

68. *Origins of the expanding Universe: 1912-1932*, M.J. Way & D. Hunter, (eds.), Astronomical Society of the Pacific Conference Proceedings, 471, (2013), [ADS: 2013ASPC..471.....W].

69. www.astro.rug.nl/~vdkruit/Beijing.html.

70. National Optical Astronomical Observatories, www.noao.edu/image_gallery/.

71. Observatories of Ohio, observatoriesofohio.org/perkins-observatory/.

72. www.prewarbuick.com/cars/83/1925-Buick-Sedan.

73. *De opvolging van W. de Sitter*, by J.K. Katgert-Merkelijn, Jaarboekje voor geschiedenis en oudheidkunde van Leiden en omstreken 1997 / Vereniging 'Oud-Leiden', 128-143 (1997). www.oudleiden.nl/pdf1/1997_12.pdf.

74. *De bouw der sterrenstelsels*, by J.H. Oort, Inaugural lecture, Hemel & Dampkring, 34, 1-16 (1936).

75. *The Kenya Expeditions of Leiden Observatory*, by J.K. Katgert-Merkelijn, Journal for the History of Astronomy 22, 267-296 (1991), [ADS: 1991JHA....22..267K].

76. *Digitized Sky Surveys, based on photographic data obtained using the 48-inch Schmidt Telescope on Palomar Mountain*, Space Telescope Science Institute, stdatu.stsci.edu/cgi-bin/dss_form.

7. The Watershed: World War II

77. *Reminiscences of astronomy in the twentieth century*. Memorie della Societa Astronomica Italiana, 53, 795-801 (1982). [ADS: 1982MmSAI..53..795O].

78. *De Leidse Universiteit 1928-1946: Vernieuwing en verzet*, by P/.J. Idenburg, Universitaire Pers Leiden, ISBN: 9-060-21425-0 (1978).

79. Private information, Willem Otterspeer,, Leiden University and Academisch Historisch Museum Leiden.

80. www.dbnl.org/tekst/clev00326no01_01/clev00326no01_01_0002.php.

81. *The Committee for the Distribution of Astronomical Literature and the Astronomical News Letters*, by B.J. Bok & V. Kourganoff, V. Vistas in Astronomy, 1, 22-25 (1955), [ADS: 1955VA....1...22B].

82. *In memoriam Professor Frank Schlesinger*, by J.H. Oort, Hemel & Dampkring 43, 27-30 (1945).

83. *Nevels rondom Nova Persei: Een boeiende geschiedenis*, by J.J. Raimond (& J.H. Oort), Hemel & Dampkring, 41, 145-163 en 200-206 (1943).

84. www.nasa.gov/multimedia/imagegallery/image_feature_1604.html.   NASA, ESA, J. Hester, A. Loll (ASU).

85. commons.wikimedia.org/wiki/File:JJL_Duyvendak.jpg.

86. www.astronomenclub.nl/.

87. *Leidse   hoogleraren   vanaf   1575,   hoogler-aren.leidenuniv.nl/search?keyword=Oort;docsPerPage=1;startDoc=4.

88. www.erfgoedleiden.nl/schatkamer/beeldbank-wo-ii; enter ' Kleibrink' in the button 'vrij zoeken'.

89. *Cosmic Static*, by G. Reber, Astrophysical Journal, 100, 279-287 (1944), [ADS: 1944ApJ...100..279R].

90. commons.wikimedia.org/wiki/File:Grote_Antenna_Wheaton.gif. Published in [89].

8. Breaking New Ground

91. *How to tell Toledo from the night sky*, by L.M. Netzer, St. Martin's Press, EAN 9-78-142724-416-1 (2014).

92. *Heren van de thee*, by Hella S. Haasse, Querido, ISBN 9-78-902143-579-4 (1992), Translated as *Tea Lords*, Granta Books, ISBN 9-78-184627-171-7.

93. *In memoriam Dr. A. de Sitter*, by J.H. Oort, Hemel & Dampkring, 44, 33.

94. *The presentation of the Bruce Gold Medal for the year 1942 to Dr. Jan H. Oort* Publications of the Astronomical Society of the Pacific, 58, 229-232 (1946) [ADS: 1946PASP...58..229.].

95. *Some phenomena connected with interstellar matter*, George Darwin Lecture by J.H. Oort, Monthly Notices of the Royal Astronomical Society, 106, 159-179 (1946) [ADS: 1946MNRAS.106..159O].

96. commons.wikimedia.org/wiki/File:HaleTelescope-MountPalomar.jpg.

97. krollermuller.nl/nl/page/15?q=keywords:John+Rädecker#filters.

9. From Kootwijk to Dwingeloo

98. *CAMRAS,* A.C. Muller Radio Astronomisch Station, uses the Dwingeloo Telescope for amateur purposes (such as 'Moon-bouncing'). Thanks to Ard Hartsuijker.

99. *Galactic structure determined from 21-cm observations,* by C.A. Muller, G. Westerhout, A. Ollongren, H.C. van de Hulst, M. Schmidt & E. Raimond, Bulletin of the Astronomical Institutes of the Netherlands, 13, 151-273 (1957), [ADS: 1957BAN....13..151M, ..196O, ..201W, ..247S, ..269R].

100. www.nobelprize.org/nomination/archive/search_people.php.

101. www.noao.edu/staff/rector/digital/rosette/index.html.

102. *The start of 21-cm line research: The early Dutch years,* by G. Westerhout, In: Seeing through the dust: The detection of HI and the exploration of the ISM in Galaxies, ASP Conference Proceedings, 276. 2002, 27-33 (2002), [2002ASPC..276...27W].

103. *De bouw van een radiotelescoop* by H. Kleibrink, (1956). www.astron.nl/about-astron/history/footage/historic-footage or www.youtube.com/watch?v=3SvkWdm-KOw.

104. *Een organisatie van en voor onderzoekers: De Nederlandse Organisatie voor Zuiver-Wetenschappelijk Onderzoek (Z.W.O.), 1947-1988,* by A.E. Kersten, Van Gorcum (1996), ISBN-13: 978-9-023-23051-9.

10. The Structure of Our Galaxy and Dwingeloo

105. *De ontdekking van de hemel,* by Harry Mulisch, ISBN 978-9-0234-2822-0 (1992). English translation *The discovery of Heaven,* ISBN 0-6-708-5668-1.

106. www.quora.com/How-is-Carl-Sagans-famous-snowflake-quote-true.

107. *A short history of nearly everything,* by Bill Bryson, ISBN 0-767-90817-1 (2003).

108. *Het bolwerk van de vrijheid, De Leidse Universiteit in heden en verleden,* by W. Otterspeer, Prometheus, ISBN 9-789-03512-2406 (2000).

109. *De vette jaren: de Commissie-Casimir en het Nederlandse wetenschapsbeleid 1957–1970,* by D.M. Baneke, Studium, 5, 110-127 (2012). www.gewina-studium.nl/articles/abstract/10.18352/studium.8195/.

110. CERN Document Service, cds.cern.ch/record/41046.

111. www.eso.org/public/images/eso0912d/.

112. en.wikipedia.org/wiki/Messier_13#/media/File:M13_from_an_8"_SCT.jpg (CC BY-SA 4.0).

113. en.wikipedia.org/wiki/Confocal_conic_sections#/media/File:Ell-hyp-konfokal.svg.

114. *The beginnings of radio astronomy in the Netherlands*, H. van Woerden & R.G. Strom, Journal of Astronomical History and Heritage, 9, 3-20 (2006), [ADS: 2006JAHH....9....3V].

115. *Dwingeloo – the golden radio telescope*, by H. van Woerden & R.G. Strom, Astronomische Nachrichten, 328, 376-387 (2007), [ADS: 2007AN....328..376V].

116. www.spacetelescope.org/images/potw1035a/.

117. *Toch nog een beetje botsen met de wolk van Gail Smith*, by Govert Schilling, de Volkskrant, January 19, 2008, pag. 5.

11. Westerbork and Retirement

118. www.astro.rug.nl/~vdkruit.

119. *The Westerbork Observatory, continuing adventure in radio astronomy*, by E. Raimond & R. Genee, Astrophysics and Space Science Library No. 208, Kluwer, ISBN 0-792-34150-3 (1996).

120. *50 Years Westerbork Radio Observatory. A continuing journey to discoveries and Innovations*, by A. van Ardenne, R.G. Strom & S. Torchinsky, online via www.astron.nl/sites/default/files/shared/Timeline/50_Years_Westerbork_Radio_Observatory.pdf.

121. *The Synthesis Radio Telescope at Westerbork, General lay-out and mechanical aspects*, J.W.M. Baars & B.G. Hooghoudt, Astronomy and Astrophysics, 31, 323 (1974) [ADS: 1974A&A....31..323B].

122. *The Synthesis Radio Telescope at Westerbork, The 21 cm continuum receiver system*, by J.L. Casse & C.A. Muller, Astronomy and Astrophysics, Vol. 31, 333 (1974) [ADS: 1974A&A....31..333C].

123. *The Synthesis Radio Telescope at Westerbork, Principles of operation, performance and data reduction*, by J.A. Högbom & W.N. Brouw, Astronomy and Astrophysics, 33, 289 (1974) [ADS: 1974A&A....33..289H].

124. *Afscheidscollege*, by J.H. Oort, Valedictory lecture, Nederlands Tijdschrift voor Natuurkunde 36, 321-325 (1970), also Hemel & Dampkring, 68, 257-161.

125. hoogleraren.leidenuniv.nl/id/1855.

12. The Last Horizon

126. *Some notes on my life as an astronomer*, by J.H. Oort, Annual Review of Astronomy and Astrophysics, 19, 1-5 (1981), [ADS: 1981ARA&A..19....1O].

127. Newspaper cutting from 'Het Parool', July 17, 1987, p. 9.

128. Newspaper cutting from 'De Telegraaf', January 11, 1986, p. T21.

129. *Horizonnen*, by J.H. Oort, Zenit, 16, 124-132 (1989).

130. *Memory of the Netherlands*, Royal National Library, translation Frits Ham. www. geheugenvannederland.nl/en/geheugen/pages/collectie/Frederik+van+\penalty-\@MEeden/De+water+lelie.

A. Astronomical Background

131. *The Hertzsprung Russell Diagram by Richard Powell*, www.atlasoftheuniverse.com/hr.html.
132. *Introductory astronomy*, by P.C. van der Kruit www.astro.rug.nl/~vdkruit/Inleiding.html.

B. Honors and Genealogy

133. ssd.jpl.nasa.gov/sbdb.cgi.
134. nl.wikipedia.org/wiki/Lijst_van_muurformules_in_Leiden.
135. commons.wikimedia.org/wiki/Category:Wall_formulas?uselang=nl#mw-subcategories.
136. 11fountains.nl/en/.

Index

© The Editor(s) (if applicable) and The Author(s), under exclusive license to Springer Nature Switzerland AG 2021

C. van der Kruit, *Master of Galactic Astronomy: A Biography of Jan Hendrik Oort*, Springer Biographies, https://doi.org/10.1007/978-3-030-55548-1

Printed in the United States
by Baker & Taylor Publisher Services